Studies in Computational Intelligence

Volume 728

Series editor

Janusz Kacprzyk, Polish Academy of Sciences, Warsaw, Poland
e-mail: kacprzyk@ibspan.waw.pl

About this Series

The series "Studies in Computational Intelligence" (SCI) publishes new developments and advances in the various areas of computational intelligence—quickly and with a high quality. The intent is to cover the theory, applications, and design methods of computational intelligence, as embedded in the fields of engineering, computer science, physics and life sciences, as well as the methodologies behind them. The series contains monographs, lecture notes and edited volumes in computational intelligence spanning the areas of neural networks, connectionist systems, genetic algorithms, evolutionary computation, artificial intelligence, cellular automata, self-organizing systems, soft computing, fuzzy systems, and hybrid intelligent systems. Of particular value to both the contributors and the readership are the short publication timeframe and the worldwide distribution, which enable both wide and rapid dissemination of research output.

More information about this series at http://www.springer.com/series/7092

Krassimir Georgiev · Michail Todorov
Ivan Georgiev
Editors

Advanced Computing in Industrial Mathematics

11th Annual Meeting of the Bulgarian Section of SIAM December 20–22, 2016, Sofia, Bulgaria. Revised Selected Papers

Editors
Krassimir Georgiev
Institute of Information and Communication Technologies
Bulgarian Academy of Science
Sofia
Bulgaria

Michail Todorov
Faculty of Applied Mathematics and Informatics
Technical University of Sofia
Sofia
Bulgaria

Ivan Georgiev
Institute of Information and Communication Technologies
Bulgarian Academy of Sciences
Sofia
Bulgaria

and

Institute of Mathematics and Informatics
Bulgarian Academy of Sciences
Sofia
Bulgaria

ISSN 1860-949X ISSN 1860-9503 (electronic)
Studies in Computational Intelligence
ISBN 978-3-319-88049-5 ISBN 978-3-319-65530-7 (eBook)
https://doi.org/10.1007/978-3-319-65530-7

Softcover reprint of the hardcover 1st edition 2017

Printed on acid-free paper

This Springer imprint is published by Springer Nature
The registered company is Springer International Publishing AG
The registered company address is: Gewerbestrasse 11, 6330 Cham, Switzerland

Preface

The 11th Annual Meeting of the Bulgarian Section of the Society for Industrial and Applied Mathematics (BGSIAM) was held in Sofia, December 20–22, 2016. The Section was formed in 2007 with the purpose to promote and support the application of mathematics to science, engineering, and technology in Bulgaria.

The goals of BGSIAM follow and creatively develop the general goals of SIAM:

- To advance the application of mathematics and computational science to engineering, industry, science, and society;
- To promote research that will lead to effective new mathematical and computational methods and techniques for science, engineering, industry, and society;
- To provide media for the exchange of information and ideas among mathematicians, engineers, and scientists.

During the BGSIAM'16 conference, a wide range of problems concerning recent achievements in the field of industrial and applied mathematics were presented and discussed. The meeting provided a forum for exchange of ideas between scientists, who develop and study mathematical methods and algorithms, and researchers, who apply them for solving real-life problems.

The topics of interest include: industrial mathematics; scientific computing; numerical methods and algorithms; hierarchical and multilevel methods; high-performance computing; partial differential equations and their applications; control and uncertain systems; Monte Carlo and quasi-Monte Carlo methods; neural networks, metaheuristics, and genetic algorithms.

The list of invited speakers include:

- Vassil Alexandrov (Barcelona Supercomputing Center, Spain), *Data and Computational Science Methods Applied to Social Media*
- Krassimir Danov (Sofia University, Bulgaria), *Modeling of Membranes with Complex Rheology: Computational Aspects*
- Oleg Iliev (Fraunhofer ITWM, Germany), *Toward MLMC Based Exascale Computations for Uncertainty Quantification for Flow in Porous Media*

- Ivan Markovsky (Vrije Universiteit Brussel, Belgium), *A Low-rank Matrix Completion Approach to Data-driven Signal Processing*

We would like to thank all the referees for the constructive remarks and criticism, which furthered considerable improvements of the quality of the papers in this book.

Sofia, Bulgaria

Krassimir Georgiev
Michail Todorov
Ivan Georgiev

Contents

Local Perturbation Analysis of the Stochastic Matrix Riccati Equation with Applications in Finance

Vera Angelova

Abstract In this paper a local perturbation analysis of the stochastic matrix Riccati equation /SMRE/ with applications in linear quadratic optimization of stochastic finance models is made. Rewriting the SMRE in equivalent form of affine linear operators and applying the techniques of Fréchet derivatives, absolute and relative norm-wise condition numbers are derived and local (first order) perturbation bounds for the error in the computed solution are formulated. The condition numbers and the perturbation bounds allow to estimate the conditioning of the SMRE and the accuracy of its computed by a numerical stable algorithm solution.

1 Introduction

The stochastic linear quadratic /SLQ/ control approach proves to give effective and appropriate solutions to investment problems in finance [1]. When applying the general SLQ control approach to study the problem of tracking a financial benchmark via trading a portfolio of a small number of assets, the homogeneous canonical form of the SLQ control model is used [1, 2] with objective

$$\min \ E\left[\int_0^\infty [x(t)^\top Qx(t) + u(t)^\top Ru(t)]dt\right] \tag{1}$$

subject to the portfolio model

$$\begin{cases} dx(t) = [Ax(t) + Bu(t)]dt + \sum_{j=1}^{k} \left[C_j x(t) + D_j u(t)\right] dW_j(t), \\ x(0) = x_0, \end{cases} \tag{2}$$

V. Angelova (✉)
Institute of Information and Communication Technologies - BAS,
Akad. G. Bonchev, Str., Bl. 2, Sofia, Bulgaria
e-mail: vangelova@iit.bas.bg

K. Georgiev et al. (eds.), *Advanced Computing in Industrial Mathematics*, Studies in Computational Intelligence 728,
https://doi.org/10.1007/978-3-319-65530-7_1

where the market index is represented as a weighted sum of the constituents, each of which is modeled by a geometric Brownian motion $W_j(t)$ and $A, C_j \in \mathbb{R}^{n\times n}$, $B, D_j \in \mathbb{R}^{n\times m}$ for $j = \overline{1,k}$. The vectors $u(t)$ and $x(t)$ are the control and the state vectors, respectively, and E denotes the mathematical expectation. The SLQ problem consists in to identify an appropriate Riccati-type equation (see [3–6] and the references therein). Then, the solvability of the SLQ problem is equivalent to the solvability of the Riccati equation.

The matrix Riccati equation subject to the SLQ problem (1), (2) is the stochastic matrix Riccati equation /SMRE/

$$F(X,P) := A^{\top}X + XA + Q + \sum_{j=1}^{k} C_j^{\top}XC_j - \\ \left(XB + \sum_{j=1}^{k} C_j^{\top}XD_j\right)\left(R + \sum_{j=1}^{k} D_j^{\top}XD_j\right)^{-1}\left(XB + \sum_{j=1}^{k} C_j^{\top}XD_j\right)^{\top}, \quad (3)$$

for an unknown real symmetric matrix $X \in \mathbb{R}^{n\times n}$, which satisfies the inequality

$$R + \sum_{j=1}^{k} D_j^{T}XD_j > 0,$$

and data collection $P \subset \mathscr{P} := \{A, B, Q, R, C_1, D_1, C_2, D_2, \dots, C_k, D_k\}$.

If the state weighting matrix Q is a positive semi-definite and the control weighting matrix R is a positive definite, the solution to (3) can be obtained by Newton iteration, Lyapunov iteration, LMI approach, as in the LMI approach the problem to solve the SMRE (3) is avoided by solving the associated with SMRE (3) convex optimization problem called semidefinite programming problem, see [1, 7, 8] and the references therein. Then, an optimal control, based on the solution to SMRE (3) is:

$$u(t) = -\left(R + \sum_{j=1}^{k} D_j^{\top}XD_j\right)^{-1}\left(B^{\top}X + \sum_{j=1}^{k} D_j^{\top}XC_j\right)x(t).$$

The problem of existence and uniqueness of the solution to SMRE (3) is considered by Rami and Zhou in [9] and then extended by Ivanov and Lomev in [8], where two numerically effective iterations are proposed and compared with the LMI approach.

In order to accurately and effectively solve Eq. (3) on a computer, a numerically stable algorithm, as well as the knowledge of the sensitivity and the conditioning of the equation to perturbations in the data are needed. A measure of the conditioning of a computational problem are its condition numbers given by the ratio of the relative changes in the solution to the relative changes in the argument. The condition numbers are involved in the formulation of perturbation upper bounds of the error in the computed solution. In turn, the perturbation error bounds estimate the sensitivity of the computational problem and are one of the elements of the high-performance computations.

Perturbation analysis for the algebraic Riccati equations, arising in stochastic control problems are made by many authors [10–13] and the references therein, while from the best of our knowledge the sensitivity of the SMRE (3) is still not analyzed.

In this paper, the conditioning and the sensitivity of the stochastic matrix Riccati Eq. (3) are studied. Norm-wise absolute and relative condition numbers are proposed. For this purpose, Eq. (3) is rewritten in equivalent form using affine linear operators. Then the techniques of Fréchet derivatives are applied. Local perturbation bounds, based on the condition numbers and neglecting terms of second and higher order are formulated as well. The local bounds are only asymptotically valid and they work even then the perturbed equation has not a unique solution in a neighborhood of the unperturbed solution.

The following notations are used later on: $\mathbb{R}^{n\times m}$ is the set of $n \times m$ matrices over the field of real numbers $\mathbb{R}$; I_n is the identity $n \times n$ matrix; $A^\top$ stands for the transpose of A; $\mathrm{vec}(A) = \left[a_1^\top, a_2^\top, \dots, a_n^\top\right]^\top \in \mathbb{R}^{n^2}$ is the column-wise vector representation of the matrix $A = \left[a_1, a_2, \dots, a_n\right] \in \mathbb{R}^{n\times n}$, $a_j \in \mathbb{R}^n$, where $\mathbb{R}^n = \mathbb{R}^{n\times 1}$; $\Pi_{n^2} \in \mathbb{R}^{n^2\times n^2}$ is the so called vec-permutation matrix such that for each $Y \in \mathbb{R}^{n\times n}$ it is fulfilled $\mathrm{vec}(Y^\top) = \Pi_{n^2}\mathrm{vec}(Y)$; $A \otimes B = [A(k,l)B]$ is the Kronecker product of the matrices $A = [A(k,l)]$ and B; $\|\cdot\|$ is the induced norm in the space of linear operators; $\|\cdot\|_2$ is the Euclidean vector or the spectral matrix norm; $\|\cdot\|_{\mathrm{F}}$ is the Frobenius norm; The notation '$:=$' stands for 'equal by definition'.

The paper is organized as follows. The problem is stated in Sect. 2. In Sect. 3 condition numbers and local perturbation bounds are derived. The paper terminates with concluding remarks in Sect. 4.

2 Statement of the Problem

When we solve a well conditioned problem $F(X,P)$ with a numerically stable iterative algorithm in finite precision arithmetic with machine precision ε, the calculated solution $\hat{X} = X + \delta X$ is the right solution to a problem in the neighborhood $F(X + \delta X, P + \delta P)$ of the problem solved $F(X,P)$.

The term δX, $\|\delta X\|_{\mathrm{F}} \le \varepsilon \|X\|_2$ reflects the presence of round-off errors and errors of approximation in the solution $\hat{X}$ computed in environment with machine precision ε. The round-off errors, the errors of approximation or data uncertainties are represented as perturbations δZ_i in the matrices $\hat{Z}_i = Z_i + \delta Z_i$ from the perturbed data collection $\hat{P} = P + \delta P \in \mathscr{P}^* := \{Z_1, Z_2, \dots, Z_r\} \subset \mathscr{P}$—the set of all matrices from $\mathscr{P}$, each of which is perturbed. If some of the above matrices are note perturbed, then the corresponding perturbations are assumed to be zero.

The local perturbation problem for SMRE (3) is to estimate norm-wisely the conditioning of (3) and to formulate a first order local bound

$$\delta_X = \|\delta X\|_F \leq f(\delta) + O(\|\delta\|^2), \ \delta \to 0$$

for the error δX in the computed solution $\hat{X}$ in terms of the perturbations δZ_i in the data matrices $\hat{Z}_i = Z_i + \delta Z_i$

$$\delta = [\delta_{Z_1}, \delta_{Z_2}, \dots \delta_{Z_r}]^\top = [\|\delta Z_1\|_F, \|\delta Z_2\|_F, \dots \|\delta Z_r\|_F]^\top$$
$$Z_i \in \mathscr{P}^* = \{Z_1, Z_2, \dots, Z_r\} \subset \mathscr{P} = \{A, B, Q, R, C_1, D_1, C_2, D_2, \dots, C_k, D_k\}$$

3 Local Perturbation Analysis

3.1 Equivalent Operator Form to SMRE (3)

Rewrite the considered SMRE (3) in equivalent form, using affine linear operators

$$F(X, P) := F_1(X, P_1) - F_2(X, P_2) F_3(X, P_3)^{-1} F_2(X, P_2)^\top = 0,$$

where the symmetric fractional affine matrix operators $F_i(X, P_i)$ are defined from

$$F_i(X, P_i) = S_i + V_i^\top X + X V_i + \sum_{j=1}^{k} Y_{ij}^\top X Y_{ij}, \quad i = 1, 3$$

with

$$S_1 = Q, \quad V_1 = A, \quad Y_{1j} = C_j$$
$$S_3 = R, \quad V_3 = 0, \quad Y_{3j} = D_j,$$

and

$$F_2(X, P_2) = XB + \sum_{j=1}^{k} C_j^\top X D_j.$$

The fractional affine matrix operators $F_i(X, P_i)$, $i = \overline{1, 3}$ depend on the matrix collections

$$P := \{P_1, P_2, P_3\}$$
$$P_1 := \{A, Q, C_1, C_2, \dots, C_k\}$$
$$P_2 := \{B, C_1, D_1, C_2, D_2, \dots, C_k, D_k\}$$
$$P_3 := \{R, 0, D_1, D_2, \dots, D_k\}.$$

3.2 Perturbed SMRE

Let the matrices Z from the data collection $\mathscr{P} = \{A, B, Q, R, C_1, D_1, C_2, D_2, \dots, C_k, D_k\}$ be perturbed with some perturbation δZ. The perturbation δZ reflects round-off errors, errors of approximation or data uncertainties.

Denote by $P + \delta P$ the perturbed data collection $\hat{P}$, in which each matrix $Z \in \mathscr{P}$ be replaced by $Z + \delta Z$. Denote by $\mathscr{P}^* := \{Z_1, Z_2, \dots, Z_r\} \subset \mathscr{P}$ the set of all matrices from $\mathscr{P}$, which are perturbed.

The perturbed SMRE is

$$F(X + \delta X, P + \delta P) = 0. \tag{4}$$

The perturbation δX in the solution $X + \delta X$ of the perturbed SMRE (4) is due to the perturbations δP in the matrix coefficients from the perturbed data collection $\mathscr{P}^*$.

3.3 Condition Numbers

Having in mind that $F(X, P) = 0$, the perturbed Eq. (4) may be written as

$$F(X + \delta X, P + \delta P) := F_X(\delta X) + \sum_{Z \in \mathscr{P}^*} F_Z(\delta Z) + G(\delta X, \delta P) = 0,$$

where the term $G(\delta X, \delta P)$ contains second and higher order terms in δX, δP and $F_Z(\delta Z) := F_Z(X, P)(\delta Z)$ are the Fréchet derivatives of $F(X, P)$ in the corresponding matrix argument $Z \in \mathscr{P}^*$ or $Z = X$, computed at the point (X, P):

$$\begin{aligned} F_X(Z) = A^\top Z + ZA + \sum_{j=1}^{k} C_j^\top Z C_j - \left(ZB + \sum_{j=1}^{k} C_j^\top Z D_j \right) N \\ - M \left(B^\top Z + \sum_{j=1}^{k} D_j^\top Z C_j \right) + M \left(\sum_{j=1}^{k} D_j^\top Z D_j \right) N, \end{aligned}$$

and

$$\begin{aligned} F_Q(Z) &:= Z \\ F_R(Z) &:= -MZN \\ F_A(Z) &:= Z^\top X + XZ \\ F_B(Z) &:= -XZN - MZ^\top X \\ F_{C_j}(Z) &:= Z^\top X C_j + C_j^\top X Z - Z^\top X D_j N - M D_j^\top X Z, \quad j = 1, \dots, m \\ F_{D_j}(Z) &:= -C_j^\top X Z N - M Z^\top X C_j + M \left(Z^\top X D_j + D_j^\top X Z \right) N, \ j = 1, \dots, m, \end{aligned}$$

with

$$N := F_3(X, P_2)^{-1} F_2(X, P_2)^\top; \quad M := F_2(X, P_2) F_3(X, P_3)^{-1}$$

The matrix representation L_Z of the operator $F_Z(.)$ is:

$$\begin{aligned} L_X &:= I \otimes A^\top + A^\top \otimes I + \sum_{j=1}^{k} C_j^\top \otimes C_j^\top - (BN)^\top \otimes I - \sum_{j=1}^{k} (D_j N)^\top \otimes C_j^\top \\ &\quad - I \otimes MB^\top - \sum_{j=1}^{k} C_j^\top \otimes MD_j^\top + \sum_{j=1}^{k} (D_j N)^\top \otimes MD_j^\top \\ &= I \otimes A^\top + A^\top \otimes I - (BN)^\top \otimes I - I \otimes MB^\top \\ &\quad + \sum_{j=1}^{k} \left(C_j^\top \otimes C_j^\top - (D_j N)^\top \otimes C_j^\top - C_j^\top \otimes MD_j^\top + (D_j N)^\top \otimes MD_j^\top \right), \end{aligned} \tag{5}$$

when $Z = X$ and

$$\begin{aligned} L_Q &= I_{n^2} \\ L_R &= -N^\top \otimes M \\ L_A &= (X^\top \otimes I) \Pi_{n^2} + I \otimes X \\ L_B &= -N^\top \otimes X - (X^\top \otimes M) \Pi_{nm} \\ L_{C_j} &= \left((XC_j)^\top \otimes I \right) \Pi_{n^2} + I \otimes C_j^\top X - \left((XD_j N)^\top \otimes I \right) \Pi_{n^2} - I \otimes MD_j^\top X \\ L_{D_j} &= -N^\top \otimes C_j^\top X - \left((XC_j)^\top \otimes M \right) \Pi_{nm} + \left((XD_j)^\top \otimes M \right) \Pi_{nm} + N^\top \otimes MD_j^\top X, \end{aligned} \tag{6}$$

when $Z \in \mathscr{P}^*$.

Assume that the SMRE (3) has a solution X, such that the linear operator $F_X(X, P)$ is invertible. This leads to the statements:

- The perturbed SMRE (4) has an unique isolated solution $\hat{X} = X + \delta X$ in the neighborhood of X for sufficiently small perturbations δP in data collection P;
- the elements of δX are analytic functions of the data perturbations δP.

Since the operator $F_X(.)$ is invertible we get

$$\delta X = -\sum_{Z \in \mathscr{P}^*} F_X^{-1} \circ F_Z(\delta Z) - F_X^{-1}(G(\delta X, \delta P)),$$

or in vector form

$$\operatorname{vec}(\delta X) = -\sum_{Z \in \mathscr{P}^*} L_X^{-1} L_Z \operatorname{vec}(\delta Z) - L_X^{-1} \operatorname{vec}(G(\delta X, \delta P)) \tag{7}$$

Hence, for the Frobenius norm of the perturbation δX in the solution X of (3) we get an absolute estimate

$$\delta_X := \|\delta X\|_{\mathrm{F}} \leq \sum_{Z \in \mathscr{P}^*} K_Z \delta_Z + O(\|\delta\|^2), \quad \delta \to 0, \tag{8}$$

where $\delta := \left[\delta_{Z_1}\ \delta_{Z_2}\ \ldots\ \delta_{Z_r}\right]^{\mathrm{T}} \in \mathbb{R}_+^r$ is the vector of non-zero absolute norm perturbations $\delta_{Z_i} = \|\delta Z_i\|_{\mathrm{F}}$ of the perturbed data matrices $Z_i \in \mathscr{P}^*$ and

$$K_Z = \|F_X^{-1} \circ F_Z\|, \quad Z \in \mathscr{P}^* \tag{9}$$

are the *absolute individual condition numbers* of SMRE (3) with respect to perturbations in the matrix coefficients $Z \in \mathscr{P}^*$.

The absolute condition numbers K_Z (9) are calculated from the expression of the matrix representation L_X (5) and L_Z (6) of the operators $F_X(.)$ and $F_Z(.)$, respectively

$$K_Z = \|L_X^{-1} L_Z\|_2, \quad Z \in \mathscr{P}^*$$

When $X \neq 0$, a relative estimate, based on the *relative condition numbers*

$$k_Z = K_Z \frac{\|Z\|_{\mathrm{F}}}{\|X\|_{\mathrm{F}}}, \quad Z \in \mathscr{P}^*$$

with respect to perturbations in the data matrices $Z \in \mathscr{P}^*$ is

$$\frac{\|\delta X\|_{\mathrm{F}}}{\|X\|_{\mathrm{F}}} \leq \sum_{Z \in \mathscr{P}^*} k_Z \frac{\delta_Z}{\|Z\|_{\mathrm{F}}} + O(\|\delta\|^2), \quad \delta \to 0 \tag{10}$$

3.4 *Non-linear First Order Homogeneous Local Bound*

Local estimates as (8) and (10), based on condition numbers may produce pessimistic results. For this purpose we define the following local first-order homogeneous norm-wise estimate, derived on the base of the vector form (7) of the perturbed Eq. (4)

$$\delta_X \leq \mathrm{est}(\delta) + O(\|\delta\|^2), \quad \delta \to 0 \tag{11}$$

$$\mathrm{est}(\delta) = \min\{\mathrm{est}_1(\delta), \mathrm{est}_2(\delta)\}$$

$$\mathrm{est}_1(\delta) = \|[L_X^{-1} L_{Z_1}, L_X^{-1} L_{Z_2}, \ldots, L_X^{-1} L_{Z_r}]\|_2 \|\delta\|_2, \quad Z_i \in \mathscr{P}^* \tag{12}$$

$$\begin{aligned} &\mathrm{est}_2(\delta) = \sqrt{\delta^\top T\,\delta}, \\ &T \text{ - } r \times r \text{ matrix with elements } \|(L_X^{-1} L_{Z_i})^\top (L_X^{-1} L_{Z_j})\|_2 \end{aligned} \tag{13}$$

A possible disadvantage of the bound proposed may be the high dimensions of the involved matrices L_X and L_Z.

3.5 *Local Component-Wise Bound*

The norm-wise perturbation bounds, formulated in Sect. 3.4 are maximally compressed, neglecting the influence of particular elements of the perturbations in the matrix coefficients on the elements of δX. The norm-wise perturbation bounds may not be a good measure for the sensitivity of the solution if there are large differences in the perturbations of different elements in the data and/or the solution. To avoid this, one may use the so called component-wise bounds. The component-wise bounds estimate the influence of the perturbations in individual elements of the data on the perturbations in the elements of the solution. Such a local component-wise bound follows directly from the vector representation (7) of the relation for the perturbation δX in the solution:

$$|\mathrm{vec}(\delta X)| \preceq \sum_{Z \in \mathscr{P}^*} |L_X^{-1} L_Z|\,|\mathrm{vec}(\delta Z)| + O(\|\delta\|^2), \quad \delta \to 0$$

The implementation of a component-wise estimate needs information about the perturbations in the components of the data, e.g. $|\mathrm{vec}(Z)| \preceq \Delta_Z, Z \in \mathscr{P}^*$, where $\Delta_Z \succeq 0$ are given vectors.

4 Concluding Remarks

In this paper, using the techniques of Fréchet derivatives, a local first-order perturbation analysis is made to the stochastic matrix Riccati Eq. (3) with applications in SLQ control of financial problems. Absolute and relative condition numbers, as well as local bounds neglecting terms of order $O(\|\delta\|^2)$ are formulated. The condition numbers and the perturbation bounds allow to estimate the conditioning of the SMRE and the accuracy of its computed by a numerical stable algorithm solution. The local bounds are valid only asymptotically, for $\delta \to 0$. Unfortunately, it is usually impossible to say, having a small but a finite perturbation δ, whether the neglected terms are indeed negligible. Moreover, for some critical values of the perturbations in the data coefficients the solution may not exist (or may go to infinity when these critical values are approached), but the local estimates will still produce a 'bound' for a very large or even for a non-existing solution. The disadvantages of the local estimates may be overcome using the techniques of non-local perturbation analysis.

Acknowledgements I am grateful to Prof. I. Ivanov for helpful suggestions.

References

1. Yao, D., Zhang, S.Z., Zhou, X.Y.: Tracking a financial benchmark using a few assets. Oper. Res. **54**(2), 232–246 (2006)
2. Zhou, X.Y., Li, D.: Continuous-time mean-variance portfolio selection: a stochastic LQ framework. Appl. Math. Optim. **42**, 19–33 (2000)
3. Huang, J., Yu, Z.: Solvability of indefinite stochastic Riccati equations and linear quadratic optimal control problems. Syst. Control Lett. **68**, 68–75 (2014)
4. Ni, Y.-H., Li, X., Zhang, J.-F.: Linear-quadratic control of discrete-time stochastic systems with indefinite weight matrices and mean-field terms. In: Preprints of the 19th World IFAC Congress Cape Town, South Africa, vol. 24–29, pp. 9750–9755 (2014)
5. Rami, M.A., Moore, J.B., Zhou, X.: Indefinite stochastic linear quadratic control and generalized differential Riccati equation. SIAM J. Control Optim. **40**, 1296–1311 (2001)
6. Wu, H., Zhou, H.Y.: Characterizing all optimal controls for an indefinite stochastic linear quadratic control problem. IEEE Trans. Autom. Control **AC-47**, 1119–1122 (2002)
7. Ivanov, I.G.: The LMI approach for stabilizing of linear stochastic systems. Int. J. Stoch. Anal. (2013). Article ID 281473, 5 pages
8. Ivanov, I.G., Lomev, B.: Numerical properties of stochastic linear quadratic model with applications in fiance. Online J. Sci. Technol. **2**(3), 41–46 (2012)
9. Rami, M., Zhou, X.: Linear matrix inequalities. Riccati equations, and indefinite stochastic linear quadratic control. IEEE Trans. Autom. Control **AC-45**, 1131–1143 (2000)
10. Chiang, C.-Y., Fan, H.-Y., Lin, M.M., Chen, H.-A.: Perturbation analysis of the stochastic algebraic Riccati equation. J. Inequal. Appl. **2013**, 580 (2013). http://www.journalofinequalitiesandapplications.com/content/2013/1/580
11. Hasanov, V.I.: Perturbation theory for linearly perturbed algebraic Riccati equations. Numer. Funct. Anal. Optim. **35**(12), 1532–1559 (2014). doi:10.1080/01630563.2014.895765
12. Ivanov, I.G., Hasanov, V.: Perturbation estimates for the two kinds of algebraic Riccati equations arising in a stochastic control. J. Num. Math. Stichastics **6**(1), 1–20 (2014)
13. Sun, J.-G.: Perturbation theory for algebraic Riccati equations. SIAM J. Matrix Anal. Appl. **19**, 39–65 (1998)

An Embedded Compact Scheme for Biharmonic Problems in Irregular Domains

Matania Ben-Artzi, Jean-Pierre Croisille and Dalia Fishelov

Abstract In Ben-Artzi et al. (SIAM J Numer Anal 47:3087–3108 (2009), [1]) a Cartesian embedded finite difference scheme for biharmonic problems has been introduced. The design of the scheme relies on a 19-dimensional polynomial space. In this paper, we show how to simplify the implementation by introducing a directional decomposition of this space. The boundary is handled via a level-set approach. Numerical results for non convex domains demonstrate the fourth order accuracy of the scheme.

1 Introduction

Let $\Omega \subseteq \mathbb{R}^2$ be a convex domain. The problem considered here is the biharmonic problem subject to Dirichlet boundary conditions:

$$\begin{cases} \Delta^2 \psi(\mathbf{x}) = f, & \mathbf{x} \in \Omega, \\ \psi = \dfrac{\partial \psi}{\partial n} = 0, & \mathbf{x} \in \partial\Omega. \end{cases} \tag{1}$$

Our purpose is to calculate a high order accurate approximation to (1), by embedding Ω in a Cartesian grid. The main idea of the scheme was described in [1]. Here we extend and elaborate on the presentation in [3, Chap. 11].

M. Ben-Artzi
Institute of Mathematics, The Hebrew University, 91904 Jerusalem, Israel
e-mail: mbartzi@math.huji.ac.il

J.-P. Croisille (✉)
Department of Mathematics, IECL, UMR CNRS 7502,
Université de Lorraine, 57045 Metz, France
e-mail: jean-pierre.croisille@univ-lorraine.fr

D. Fishelov
Afeka Tel Aviv Academic College of Engineering, 218 Bnei-Efraim St.,
69107 Tel-Aviv, Israel
e-mail: daliaf@afeka.ac.il

K. Georgiev et al. (eds.), *Advanced Computing in Industrial Mathematics*, Studies in Computational Intelligence 728,
https://doi.org/10.1007/978-3-319-65530-7_2

We consider the convex domain Ω as embedded in a large uniform grid of mesh size h. A grid point is a point $Q_{i,j} = (ih, jh)$ for $i, j \in \mathbb{Z}$. Following common terminology, we use the term **interior nodes** for the grid points that lie **inside** Ω. We denote by Ω_h the ensemble of these nodes, namely:

$$\Omega_h = \left\{ Q_{i,j} \in \Omega, \quad i,j \in \mathbb{Z} \right\}. \tag{2}$$

We split the set Ω_h into two sets, $\Omega_h = \Omega_h^{calc} \cup \Omega_h^{edge}$, as follows:

- Ω_h^{calc} = the set of *calculated nodes*.
  This set consists of those nodes that are located "well within" Ω, namely sufficiently far from the boundary $\partial\Omega$. In particular, if all diagonally neighboring nodes $Q_{i\pm1,j\pm1}$ are in Ω_h then $Q_{i,j} \in \Omega_h^{calc}$. Remark that by convexity all eight neighboring nodes are then in Ω_h. However, it should be emphasized that even if not all its neighboring nodes are in Ω_h, a node $Q_{i,j}$ can still be considered as "calculated" if it is not "too close" to the boundary, as we explain below.
  The approximate values at the calculated nodes are obtained by the proposed scheme.
- Ω_h^{edge} = the set of *edge nodes*.
  This set consists of those nodes (interior to Ω) that are located "too close" to the boundary $\partial\Omega$. They differ from the calculated nodes in the sense that there are no approximate values associated with them. Their role is "geometric"; they serve in the determination of a set Ω_h^{bdry} of **boundary nodes** that are actually located **on the boundary** $\partial\Omega$, and carry the assigned boundary values.
- Observe that the set Ω_h^{bdry} consists of selected points on the boundary, and in general is not included in the underlying global grid $Q_{i,j}$, $i, j \in \mathbb{Z}$.

In Fig. 1 we designate the calculated nodes with black circles, whereas the edge nodes are designated by white circles.

The proposed scheme is a compact scheme, i.e. all approximate values of high order derivatives are related to values of a function ψ and its derivatives ψ_x, ψ_y at immediate neighbors. More specifically, given a node $\mathbf{M}_0 = Q_{i,j} \in \Omega_h$, we consider the eight surrounding nodes in the grid:

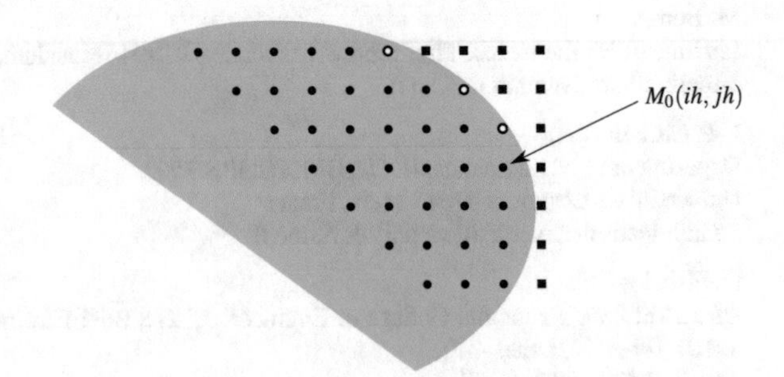

Fig. 1 Embedding of an elliptical domain in a Cartesian grid. Calculated nodes are represented by *black circles*. Exterior points are represented by *black squares*. The points labelled with *white circles* represent edge points, i.e. interior points close to the boundary

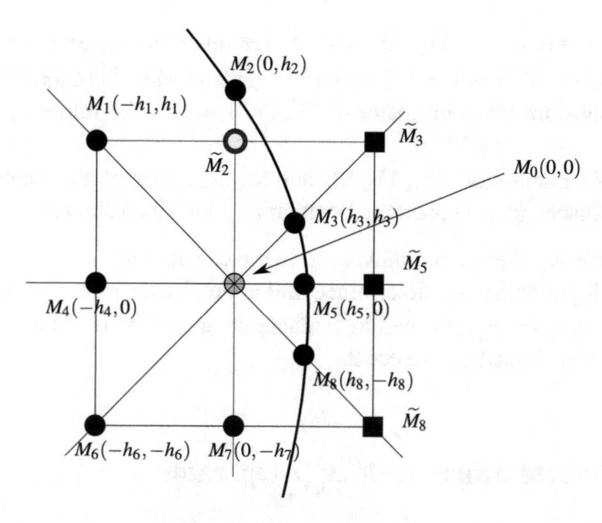

Fig. 2 Zoom on the neighborhood of point M_0 in Fig. 1. The coordinates have been moved such that M_0 is the coordinates *center*. The 8 neighbors points of M_0 are the points $\mathbf{M}_1$, $\mathbf{M}_2$, $\mathbf{M}_3$, $\mathbf{M}_4$, $\mathbf{M}_5$, $\mathbf{M}_6$, $\mathbf{M}_7$ and $\mathbf{M}_8$. The points $\mathbf{M}_1$, $\mathbf{M}_4$, $\mathbf{M}_6$ and $\mathbf{M}_7$ belong to the Cartesian grid. The points $\mathbf{M}_2$, $\mathbf{M}_3$, $\mathbf{M}_5$ and $\mathbf{M}_8$ belong to the boundary of the domain. They are obtained as the intersection of rays emanating from $\mathbf{M}_0$ and directed towards $\widetilde{\mathbf{M}}_2$, $\widetilde{\mathbf{M}}_3$, $\widetilde{\mathbf{M}}_5$ and $\widetilde{\mathbf{M}}_8$ respectively. The points $\widetilde{\mathbf{M}}_3$, $\widetilde{\mathbf{M}}_5$ and $\widetilde{\mathbf{M}}_8$ are outside the domain. The edge point above $\mathbf{M}_0$ is marked with an *open circle*

$$\widetilde{\mathbf{M}}_1 = Q_{i-1,j+1},\ \ \widetilde{\mathbf{M}}_2 = Q_{i,j+1},\ \ \widetilde{\mathbf{M}}_3 = Q_{i+1,j+1}, \widetilde{\mathbf{M}}_4 = Q_{i-1,j},$$
$$\widetilde{\mathbf{M}}_5 = Q_{i+1,j},\ \ \widetilde{\mathbf{M}}_6 = Q_{i-1,j-1},\ \ \widetilde{\mathbf{M}}_7 = Q_{i,j-1}, \widetilde{\mathbf{M}}_8 = Q_{i+1,j-1}.$$

If all the nine nodes $\widetilde{\mathbf{M}}_i$ are *calculated nodes*, namely, in Ω_h^{calc}, or coincide with a boundary point, which is part of the grid, we set $\mathbf{M}_i = \widetilde{\mathbf{M}}_i$, $i = 0, \ldots, 8$, and continue with this regular stencil centered at $\mathbf{M}_0$. Otherwise, our goal is to replace the $\widetilde{\mathbf{M}}_i' s$ that are not in Ω_h^{calc} by suitable $\mathbf{M}_i' s$ that are boundary points, namely, in Ω_h^{bdry}. The values of ψ, ψ_x, ψ_y at these points are all that is needed in order to calculate the various approximate derivatives at $\mathbf{M}_0$.

To describe this construction, suppose that $\mathbf{M}_0 \in \Omega_h^{calc}$ is a calculated node, while for some $1 \leq i \leq 8$, the neighboring node $\widetilde{\mathbf{M}}_i$ is either an edge node or an exterior node. Consider the calculated node designated by $\mathbf{M_0}$ in Fig. 1. A zoom is shown on Fig. 2. The 8 points $\widetilde{\mathbf{M}}_i$ are the points in the square (4 corner points and 4 mid-edge points). Take the ray that emanates from $\mathbf{M}_0$ and goes through $\widetilde{\mathbf{M}}_i$. This ray must cross the boundary $\partial\Omega$ at exactly one point since Ω is convex. We define the intersection point as $\mathbf{M_i}$.

The calculation of the approximate value to $\Delta^2\psi(\mathbf{M_0})$ relies on the data at $\mathbf{M_i}$ rather than $\widetilde{\mathbf{M}}_i$.

- The four neighbors $\widetilde{\mathbf{M}}_1$, $\widetilde{\mathbf{M}}_4$, $\widetilde{\mathbf{M}}_6$ and $\widetilde{\mathbf{M}}_7$ are other calculated nodes so we keep them, i.e. $\widetilde{\mathbf{M}}_i = \mathbf{M}_i$, $i = 1, 4, 6, 7$. In particular, if we shift the coordinates of $\mathbf{M}_0$ to $(0, 0)$, we have for the coordinates of $\mathbf{M}_i$, $i = 1, 4, 6, 7$, the values $h_1 = h_4 = h_6 = h_7 = h$.
- The other four neighbors $\widetilde{\mathbf{M}}_2$, $\widetilde{\mathbf{M}}_3$, $\widetilde{\mathbf{M}}_5$ and $\widetilde{\mathbf{M}}_8$ are either edge or exterior nodes so they are replaced by points on the boundary as described above.

We thus obtain $\mathbf{M}_i$, the actual points used in the calculation.

Once the 8 points $\mathbf{M}_i$ are determined and approximate values ψ, ψ_x and ψ_y are assigned to them, we can proceed to evaluate an approximate value for $\Delta^2\psi$ at the point $\mathbf{M}_0$. This is described in Sect. 2.

2 The Discrete Biharmonic $\Delta_{\mathbf{h}}^2\psi$ Operator

In this section we present our finite-difference scheme for the approximation of the biharmonic operator. Figure 2 shows the stencil used for the approximation of $\Delta^2\psi$ at $\mathbf{M}_0 = (0, 0)$. The 8 points $\mathbf{M}_k$, $1 \leq k \leq 8$ form an irregular stencil around $\mathbf{M}_0$. Each of the nine grid points $\mathbf{M}_k$ carries three values: ψ, ψ_x, ψ_y. These are calculated values if $\mathbf{M}_k \in \Omega_h^{calc}$ is a calculated node. If $\mathbf{M}_k \in \Omega_h^{bdry}$, then this point carries boundary data given by the boundary conditions. In order to approximate $\Delta^2\psi$ of a given smooth function ψ at $\mathbf{M_0}$ we interpolate the data ψ, ψ_x, ψ_y on the stencil $\{\mathbf{M}_0, \ldots, \mathbf{M}_8\}$ by a certain polynomial $P_{\mathbf{M}_0}$ of degree 6. The detailed construction of $P_{\mathbf{M}_0}(x, y)$ is carried out in Sect. 3. To handle the irregular stencil around $\mathbf{M}_0$ we denote by $\mathbf{h}$ the vector of the step-sizes around $\mathbf{M}_0$, as in Fig. 2:

$$\mathbf{h} = [h_1, \ldots, h_8]^T. \tag{3}$$

Once the polynomial $P_{\mathbf{M}_0}(x, y)$ is constructed, we replace the smooth function ψ by a discrete function $\widetilde{\psi}$, defined only on the set of nodes $\Omega_h^{calc} \cup \Omega_h^{bdry}$. The discrete biharmonic operator $\Delta_{\mathbf{h}}^2\psi$ for the approximation of $\Delta^2\psi$ at $\mathbf{M}_0 = (0, 0)$ is then defined by

$$\Delta_{\mathbf{h}}^2\widetilde{\psi}(\mathbf{M_0}) = \Delta^2 P_{\mathbf{M}_0}(0, 0), \tag{4}$$

3 Calculating the Interpolation Polynomial $P_{\mathbf{M_0}}(x, y)$

As mentioned above, our compact scheme for the biharmonic problem relies on an interpolation polynomial of degree six. Such a polynomial is constructed for every calculated point $\mathbf{M}_{i,j} \in \Omega_h^{calc}$. This sixth-order polynomial is called $P_{\mathbf{M_0}}(x, y)$. It is of the form (where here and below the subscript $\mathbf{M}_0$ is omitted),

$$P(x,y) = \sum_{i=1}^{19} a_i l_i(x,y), \tag{5}$$

where the polynomials $l_i(x,y)$ are (x,y) are shifted so that $\mathbf{M}_0 = (0,0)$:

$$\begin{cases} l_1(x,y) = 1, \;\; l_2(x,y) = x, \;\; l_3(x,y) = x^2, \;\; l_4(x,y) = x^3, \\ l_5(x,y) = x^4, \;\; l_6(x,y) = x^5, \;\; l_7(x,y) = y, \;\; l_8(x,y) = y^2, \;\; l_9(x,y) = y^3, \\ l_{10}(x,y) = y^4, \;\; l_{11}(x,y) = y^5, \;\; l_{12}(x,y) = xy, \\ l_{13}(x,y) = xy(x+y), \;\; l_{14}(x,y) = xy(x-y), \\ l_{15}(x,y) = xy(x+y)^2, \;\; l_{16}(x,y) = xy(x-y)^2, \\ l_{17}(x,y) = xy(x+y)^3, \;\; l_{18}(x,y) = xy(x-y)^3, \\ l_{19}(x,y) = x^2y^2(x^2+y^2). \end{cases} \tag{6}$$

The 19 coefficients a_i are obtained as follows. We consider the discrete values depending on $\widetilde{\psi}$ located at the eight points $\mathbf{M}_k$, $1 \le k \le 8$, around the point $\mathbf{M}_0$, (see Fig. 2). From the discrete data at these points we determine 19 values to be interpolated by $P(x,y)$ and its derivatives:

$$\begin{cases} \Gamma_1(\psi) = \widetilde{\psi}(\mathbf{M}_1), \;\; \Gamma_2(\psi) = \widetilde{\psi}(\mathbf{M}_2), \;\; \Gamma_3(\psi) = \widetilde{\psi}(\mathbf{M}_3), \\ \Gamma_4(\psi) = \widetilde{\psi}(\mathbf{M}_4), \;\; \Gamma_5(\psi) = \widetilde{\psi}(\mathbf{M}_0), \;\; \Gamma_6(\psi) = \widetilde{\psi}(\mathbf{M}_5), \\ \Gamma_7(\psi) = \widetilde{\psi}(\mathbf{M}_6), \;\; \Gamma_8(\psi) = \widetilde{\psi}(\mathbf{M}_7), \;\; \Gamma_9(\psi) = \widetilde{\psi}(\mathbf{M}_8), \\ \Gamma_{10}(\psi) = (-\partial_x + \partial_y)\widetilde{\psi}(\mathbf{M}_1), \;\; \Gamma_{11}(\psi) = \partial_y\widetilde{\psi}(\mathbf{M}_2), \\ \Gamma_{12}(\psi) = (\partial_x + \partial_y)\widetilde{\psi}(\mathbf{M}_3), \;\; \Gamma_{13}(\psi) = -\partial_x\widetilde{\psi}(\mathbf{M}_4), \\ \Gamma_{14}(\psi) = \partial_x\widetilde{\psi}(\mathbf{M}_0), \;\; \Gamma_{15}(\psi) = \partial_y\widetilde{\psi}(\mathbf{M}_0), \\ \Gamma_{16}(\psi) = \partial_x\widetilde{\psi}(\mathbf{M}_5), \;\; \Gamma_{17}(\psi) = (-\partial_x - \partial_y)\widetilde{\psi}(\mathbf{M}_6), \\ \Gamma_{18}(\psi) = -\partial_y\widetilde{\psi}(\mathbf{M}_7), \;\; \Gamma_{19}(\psi) = (\partial_x - \partial_y)\widetilde{\psi}(\mathbf{M}_8). \end{cases} \tag{7}$$

Note that the derivatives at any point are taken in the direction of $\mathbf{M}_0$ except that the full gradient is given at the point $\mathbf{M}_0$.

There is a one-to-one correspondence between the polynomial (5) and the above set of 19 data. More explicitly, the 19 coefficients a_i in (5) are uniquely determined by the data (7). For the proof of this linear algebraic fact, see [1].

In (5), the coefficients a_i depend linearly on the data $\Gamma_k(\psi), 1 \le k \le 19$. Therefore, $P(x,y)$ can be rewritten as

$$P(x,y) = \sum_{i=1}^{19} \left(\sum_{j=1}^{19} A_{i,j} \Gamma_j(\psi) \right) l_i(x,y). \tag{8}$$

We need to calculate the geometric coefficients $A_{i,j}$, $1 \leq i,j \leq 19$ in terms of the vector $\mathbf{h} = [h_1, h_2, h_3, h_4, h_5, h_6, h_7, h_8]$. For this purpose, it is useful to decompose the polynomial $P(x,y)$ into the sum of four terms

$$P(x,y) = P(0,0) + P_1(x) + P_2(y) + xyQ(x,y). \tag{9}$$

Looking at (5) and (6), these four terms are expressed as:

$$a_1 = P(0,0) = \psi(\mathbf{M}_0) \text{ (given value)}, \tag{10}$$

$$\begin{cases} P_1(x) = a_2 x + a_3 x^2 + a_4 x^3 + a_5 x^4 + a_6 x^5, \\ P_2(y) = a_7 y + a_8 y^2 + a_9 y^3 + a_{10} y^4 + a_{11} y^5. \end{cases} \tag{11}$$

The polynomial $Q(x,y)$ in (9) is then defined as

$$\begin{aligned} Q(x,y) &= \frac{P(x,y) - P(0,0) - P_1(x) - P_2(y)}{xy} \\ &= a_{12} + a_{13}(x+y) + a_{14}(x-y) + a_{15}(x+y)^2 + a_{16}(x-y)^2 \qquad (12) \\ &\quad + a_{17}(x+y)^3 + a_{18}(x-y)^3 + a_{19} xy(x^2+y^2). \qquad (13) \end{aligned}$$

This decomposition is directional in the following sense:

- The polynomial $P_1(x) \in \text{Span}\{x, x^2, x^3, x^4, x^5\}$ corresponds to the "horizontal data". It is determined by the 5 data (see Fig. 2):

$$\psi(\mathbf{M}_4), \psi(\mathbf{M}_5), \partial_x \psi(\mathbf{M}_4), \partial_x \psi(\mathbf{M}_0), \partial_x \psi(\mathbf{M}_5). \tag{14}$$

- Similarly, $P_2(y) \in \text{Span}\{y, y^2, y^3, y^4, y^5\}$ corresponds to the "vertical data". It is specified by the 5 data

$$\psi(\mathbf{M}_7), \psi(\mathbf{M}_2), \partial_y \psi(\mathbf{M}_7), \partial_y \psi(\mathbf{M}_0), \partial_y \psi(\mathbf{M}_2). \tag{15}$$

- Finally, it can be shown that the polynomial $Q(x,y)$ is determined by the 8 "diagonal data" in (7). These data are:

$$\begin{cases} \Gamma_1(\psi) = \psi(\mathbf{M_1}), \quad \Gamma_3(\psi) = \psi(\mathbf{M_3}), \quad \Gamma_7(\psi) = \psi(\mathbf{M_6}), \quad \Gamma_9(\psi) = \psi(\mathbf{M_8}), \\ \Gamma_{10}(\psi) = (-\partial_x + \partial_y)\psi(\mathbf{M_1}), \quad \Gamma_{12}(\psi) = (\partial_x + \partial y)\psi(\mathbf{M_3}), \\ \Gamma_{17}(\psi) = (-\partial_x - \partial_y)\psi(\mathbf{M_6}), \quad \Gamma_{19}(\psi) = (\partial_x - \partial_y)\psi(\mathbf{M_8}). \end{cases} \tag{16}$$

4 The Numerical Scheme

4.1 The Embedded Discrete Biharmonic Operator

In this section, we assume given for each point of the Cartesian grid the polynomial $P(x, y)$ (5) in terms of the data $\Gamma_k(\psi)$. As explained in Sect. 3, the polynomial $P_{\mathbf{M}_0}(x, y)$ in (5) is explicitly known by the coefficients a_i, given as the analytical functions:

$$\Big[\mathbf{h}, [\Gamma_j(\psi)]\Big]_{j=1,\dots,19} \mapsto \mathbf{a} = \Big[a_1, a_2, \dots, a_{18}, a_{19}\Big]^T. \tag{17}$$

The discrete biharmonic at $\mathbf{M_0}(\mathbf{x_0}, \mathbf{y_0})$ is obtained by:

$$\Delta_{\mathbf{h}}^2 \widetilde{\psi}(\mathbf{M}_0) = \sum_{k=1}^{19} a_k \Delta^2 l_k(x_0, y_0). \tag{18}$$

There are four nonvanishing terms in the right-hand-side of (18) which are:

$$\begin{cases} \Delta^2 l_5(x_0, y_0) = 24, \;\; \Delta^2 l_{10}(x_0, y_0) = 24, \\ \Delta^2 l_{15}(x_0, y_0) = 16, \;\; \Delta^2 l_{16}(x_0, y_0) = -16. \end{cases} \tag{19}$$

Therefore the discrete biharmonic at $\mathbf{M}_0$ is given in terms of the coefficients $a_k\Big[\mathbf{h}, [\Gamma_j(\psi)]\Big]$ by

$$\Delta_{\mathbf{h}}^2 \psi(M_0) \triangleq 24\Big(a_5(\mathbf{h}, [\Gamma_k(\psi)]) + a_{10}(\mathbf{h}, [\Gamma_k(\psi)])\Big) \tag{20}$$

$$+ 16\Big(a_{15}(\mathbf{h}, [\Gamma_k(\psi)]) - a_{16}(\mathbf{h}, [\Gamma_k(\psi)])\Big). \tag{21}$$

The discrete equation at point $\mathbf{M}_0$ is therefore (see (4)):

$$\Delta_{\mathbf{h}}^2 \widetilde{\psi}(\mathbf{M_0}) = f(\mathbf{M_0}). \tag{22}$$

Equation (22) has to be supplemented by some additional relation connecting the derivatives $\psi_{x,i,j}$, $\psi_{y,i,j}$ and the values $\psi_{i,j}$. Our choice [1, 3] is to use an *Hermitian* relation in the x- and the y-direction. In the x-direction we have:

$$\alpha_{1,i}\psi_{x,i-1,j} + \psi_{x,i,j} + \alpha_{2,i}\psi_{x,i+1,j} = \beta_{1,i}\psi_{i-1,j} + \beta_{2,i}\psi_{i,j} + \beta_{3,i}\psi_{i+1,j}. \tag{23}$$

The five coefficients $\alpha_{1,i}$, $\alpha_{2,i}$, $\beta_{1,i}$, $\beta_{2,i}$ and $\beta_{3,i}$ are defined as follows. Let $\mathbf{M_0} = Q_{i,j}(x_i, y_j)$ and let the two neighbor points $\mathbf{M_4}$ and $\mathbf{M_5}$ be (see Fig. 2):

$$\mathbf{M_4}(x_i - h_i, y_j), \quad \mathbf{M_5}(x_i + h_{i+1}, y_j). \tag{24}$$

Then

$$\begin{cases} \alpha_{1,i} = \frac{h_{i+1}^2}{(h_{i+1}+h_i)^2}, \quad \alpha_{2,i} = \frac{h_i^2}{(h_{i+1}+h_i)^2}, \quad \beta_{2,i} = \frac{2h_{i+1}^4+4h_{i+1}^3h_i-4h_{i+1}h_i^3-2h_i^4}{h_{i+1}(h_{i+1}+h_i)^3h_i}, \\ \beta_{1,i} = -\frac{2h_{i+1}^4+4h_{i+1}^3h_i}{h_{i+1}(h_{i+1}+h_i)^3h_i}, \quad \beta_{3,i} = \frac{2h_i^4+4h_{i+1}h_i^3}{h_{i+1}(h_{i+1}+h_i)^3h_i}. \end{cases} \tag{25}$$

In the y-direction we have

$$\gamma_{1,j}\Psi_{y,i,j-1} + \Psi_{y,i,j} + \gamma_{2,j}\Psi_{y,i,j+1} = \delta_{1,j}\Psi_{i,j-1} + \delta_{2,j}\Psi_{i,j} + \delta_{3,j}\Psi_{i,j+1}. \tag{26}$$

with values of the five coefficients $\gamma_{1,j}, \gamma_{2,j}, \delta_{1,j}, \delta_{2,j}$ and $\delta_{3,j}$ deduced from the points $\mathbf{M_7}$ and $\mathbf{M_2}$ in a way similar to (25). We refer to [1, 3] for an analysis of the Hermitian relations (23) and (26).

4.2 *Assembling the Global Linear System*

To each point (i,j) corresponds the discrete biharmonic relation (22) together with the horizontal and vertical Hermitian relations for the discrete gradient (23) and (26). All these relations form a linear system

$$A\Psi = F. \tag{27}$$

Assembling the matrix A using the relations (22), (23), (26) is analogous to assembling the global matrix in the finite element method.

According to Sect. 1, each point $\mathbf{M}_{i,j}$ of the Cartesian grid belongs to one of the five categories:

1. interior regular calculated point
2. interior irregular calculated point
3. interior edge point
4. boundary point
5. exterior point.

In our computation, this classification is performed using a so-called *level set* model for the boundary $\partial\Omega$. Assume that $(x,y) \mapsto \varphi(x,y)$ is a smooth function such that, at least locally

$$\varphi(x,y) \begin{cases} < 0 & \text{if } (x,y) \in \Omega, \quad \text{(interior point)}, \\ > 0 & \text{if } (x,y) \in \Omega^c, \quad \text{(exterior point)}, \\ = 0 & \text{if } (x,y) \in \partial\Omega, \quad \text{(boundary point)}. \end{cases} \tag{28}$$

Following [4], the interior point $\mathbf{M}_0 = \mathbf{M}_{i,j}$ is declared *close to* $\partial\Omega$ if $\varphi_{\min,i,j}\varphi_{\max,i,j} < 0$ where

$$\begin{cases} \varphi_{\min,i,j} = & \min(\varphi_{i-1,j}, \varphi_{i+1,j}, \varphi_{i,j+1}, \varphi_{i,j-1}, \varphi_{i,j}), \\ \varphi_{\max,i,j} = & \max(\varphi_{i-1,j}, \varphi_{i+1,j}, \varphi_{i,j+1}, \varphi_{i,j-1}, \varphi_{i,j}). \end{cases} \tag{29}$$

In this case, the following quadratic model for φ is defined around $\mathbf{M}_0$ by:

$$\varphi(\mathbf{x}) = \varphi_0 + (\nabla\varphi_0)^T.(\mathbf{x} - \mathbf{x}_0) + \frac{1}{2}(\mathbf{x} - \mathbf{x}_0)^T(D^2\varphi_0)(\mathbf{x} - \mathbf{x}_0). \tag{30}$$

In (30), $\nabla\varphi_0$ and $D^2\varphi_0$ stand for approximate values of the gradient and the Hessian of $\varphi(\mathbf{x})$ at $\mathbf{M}_0$. In the computations, centered differences for $\nabla\varphi_0$ and $D^2\varphi_0$ are used. Using the model (30) allows to determine the approximate projection $\mathbf{M}_0^*$ of the interior point $\mathbf{M}_0$ on $\partial\Omega$, [4]. This gives

$$\mathbf{M}_0 = \begin{cases} \text{calculated point} & \text{if } \mathrm{dist}(\mathbf{M}_0, \mathbf{M}_0^*) \geq \varepsilon_{edge}, \\ \text{edge point} & \text{if } \mathrm{dist}(\mathbf{M}_0, \mathbf{M}_0^*) < \varepsilon_{edge}. \end{cases} \tag{31}$$

where ε_{edge} is a fixed parameter. For each calculated point $\mathbf{M}_0$, the length vector $\mathbf{h} \in \mathbb{R}^8$ and the elementary matrix $A_{i,j}(\mathbf{h}) \in \mathbb{M}_{19}(\mathbb{R})$ are evaluated according to the preceding classification into regular/irregular calculated points. Finally the elements of each matrix $A_{i,j}(\mathbf{h})$ are collected in the global matrix A. In a second step, for each point $\mathbf{M}_{i,j}$, the submatrix of A corresponding to the Hermitian relations for the derivatives ψ_x and ψ_y in (23) is calculated. The global linear system $A\psi = b$ is the discrete version of the problem (1). Note that it is solved by a direct solver. Fast solvers issues in the fashion of [2, 4] will be addressed in a future work.

5 Numerical Results

We present several numerical results for the biharmonic problem with additional Laplacian term:

$$\begin{cases} \alpha\Delta^2\psi(\mathbf{x}) - \beta\Delta\psi(\mathbf{x}) = f, & \mathbf{x} \in \Omega, \\ \psi = g_1(\mathbf{x}), \quad \frac{\partial\psi}{\partial n} = g_2(\mathbf{x}), & \mathbf{x} \in \partial\Omega. \end{cases} \tag{32}$$

In each case, the domain Ω and the solution $\psi(\mathbf{x})$ are specified. The right-hand side $f(\mathbf{x})$ and the two boundary functions $g_1(\mathbf{x})$ and $g_2(\mathbf{x})$ are determined accordingly. The numerical scheme is then used to obtain an approximation for ψ based on the discrete values of f.

Table 1 Compact scheme for $\Delta^2\psi = f$. The solution is $\psi(x,y) = (1-x^2)^2(1-y^2)^2$ in the ellipse $x^2/1^2 + y^2/2^2 \leq 1$. The ellipse parameters are $(a = 1, b = 2, r = 1)$. The ellipse is embedded in the square $[-2,2] \times [-2,2]$. We present e and e_x, the l_2 errors for the stream function and for $\partial_x\psi$. The parameter for points labelled as *edge points* is $\varepsilon_{edge} = 5.10^{-3}h$

mesh	9×9	Rate	17×17	Rate	33×33	Rate	65×65
e_∞	1.1175(−2)	4.40	5.3108(−4)	3.94	3.4538(−5)	3.45	3.1596(−6)
$(e_x)_\infty$	2.3270(−2)	4.35	1.1419(−3)	3.61	9.3285(−5)	4.24	4.9262(−6)
e_2	1.7466(−2)	4.85	6.0551(−4)	4.08	3.5825(−5)	3.59	2.9702(−6)
$(e_x)_2$	3.1922(−2)	4.81	1.1402(−3)	3.79	8.2220(−5)	3.81	5.8612(−6)

Table 2 Compact scheme for $(\frac{1}{2}\Delta - \Delta^2)\psi = f$. The solution is $\psi(x,y) = 100(x^3 \ln(1+y)) + \frac{y}{1+x}$ in the ellipse $(x-0.5)^2/(0.5)^2 + (y-0.5)^2/0.3^2 \leq 1$. The ellipse parameters are $(a = 0.5, b = 0.3, r = 1)$ with center $(x_c, y_c) = (0.5, 0.5)$. The ellipse is embedded in the square $[0,1] \times [0,1]$. We present e and e_x, the l_2 errors for the stream function and for $\partial_x\psi$. The parameter for points labelled as *edge points* is $\varepsilon_{edge} = 5.10^{-3}h$

mesh	17×17	Rate	33×33	Rate	65×65	Rate	129×129
e_∞	6.9555(−6)	3.53	6.0000(−7)	4.43	2.7790(−8)	3.29	2.8334(-9)
$(e_x)_\infty$	4.0042(−4)	3.64	3.2033(−5)	4.07	1.9102(−6)	2.98	2.4215(−7)
e_2	1.1759(−6)	3.15	1.3240(−7)	4.26	6.9034(−9)	3.44	6.3715(−10)
$(e_x)_2$	7.4850(−5)	3.79	5.3933(−6)	3.98	3.4163(−7)	3.90	2.2865(−8)

5.1 *Test Cases in an Ellipse*

We first consider two test cases where the computational domain is an ellipse. A similar test case has already been considered in [1]. The observed accuracy is very good. The order of convergence is located approximately in the interval $I = [3,4]$ (Tables 1 and 2).

5.2 *Test Cases in Non Convex Domains*

5.2.1 Star Shaped Domains

We consider first the biharmonic problem (see Example 4.3 in [4])

$$\begin{cases} \Delta^2\psi(\mathbf{x}) = 0 \quad \mathbf{x} \in \Omega, \\ \psi = g_1(\mathbf{x}), \quad \frac{\partial\psi}{\partial n} = g_2(\mathbf{x}), \quad \mathbf{x} \in \partial\Omega. \end{cases} \tag{33}$$

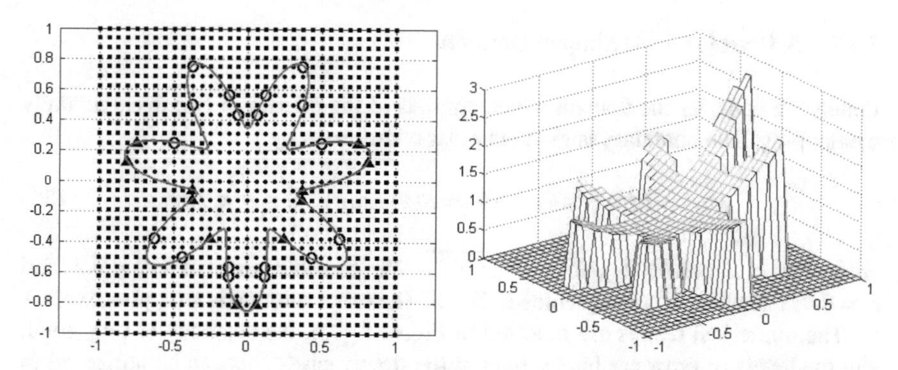

Fig. 3 Seven branches *star shaped* domain embedded in a 33×33 grid. • *Left* domain and grid. The boundary points are marked with *black triangles*. The edge points are marked with *open circles*. • *Right* approximate solution corresponding to $\psi_{ex}(x, y) = x^2 + y^2 + e^x \cos(y)$

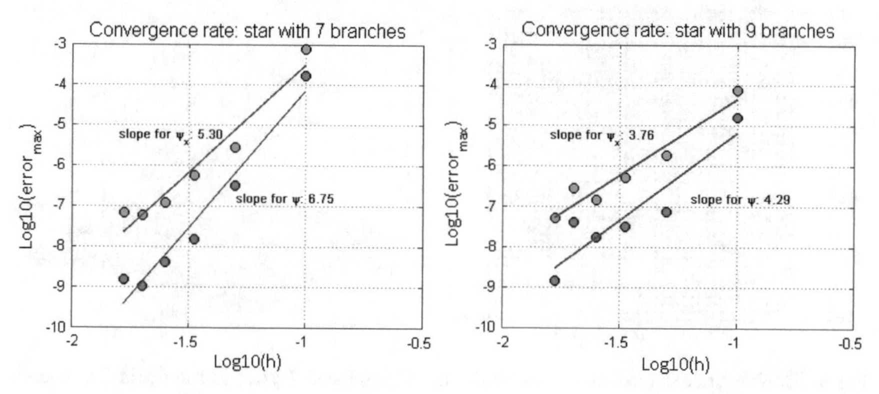

Fig. 4 Star shaped domain: linear regression of the convergence rate for $\|\Psi - \widetilde{(\psi_{ex})}\|_\infty$ and $\|\Psi_x - \widetilde{(\psi_{x,ex})}\|_\infty$ where the exact solution is $\psi_{ex}(x, y) = x^2 + y^2 + e^x \cos(y)$. • *Left* domain with 7 branches, $(k_p = 7)$. • *Right* domain with 9 branches, $(k_p = 9)$. On each regression line, the six points correspond to the six grids 10×10, 20×20, 30×30, 40×40, 50×50 and 60×60

The boundary of the domain is given in polar coordinates by

$$x(\theta) = R(\theta)\cos(\theta), \quad y(\theta) = R(\theta)\sin(\theta), \quad 0 \le \theta < 2\pi, \tag{34}$$

with $R(\theta) = 0.6 + 0.25\sin(k_p\theta)$. The domain is represented in Fig. 3 for $k_p = 7$, (seven branches case). The exact solution is $\psi(x, y) = x^2 + y^2 + e^x \cos(y)$. The numerical results are reported in Fig. 4 where the least square slope is represented, based on six grids. They show excellent accuracy, even for very coarse grids. Observe in addition the low error level for ψ and $\partial_x \psi$.

5.2.2 A Double Circle Shaped Domain

Finally we consider the domain which consists of the interior of two disks partially overlapping. The boundary is given in polar coordinates by

$$x(\theta) = R(\theta)\cos(\theta), \quad y(\theta) = R(\theta)\sin(\theta), \quad 0 \le \theta < 2\pi. \tag{35}$$

with $R(\theta) = d|\cos(\theta)| + \sqrt{R^2 - d^2 \sin(\theta)^2}$. We consider the case $R = 0.5$ and $d = 0.4$. The domain is represented in Fig. 5. The exact solution is $\psi(x, y) = \exp(x + y)$. The numerical results are reported in Fig. 6. Again, the accuracy is very good. But the levels of error are higher than in the flower case. This can be attributed to the non regular boundary.

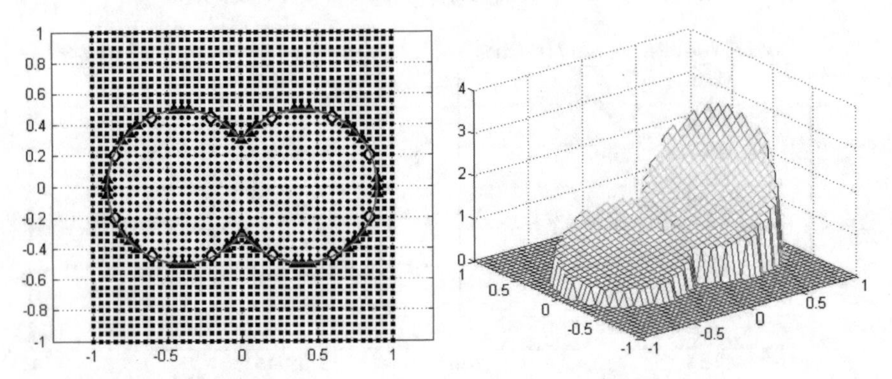

Fig. 5 Double circle shaped domain embedded in a 41×41 grid. *Left* domain and grid. The boundary points are marked with *black triangles*. The edge points are marked with *open cicles*. *Right* approximate solution $\psi(x, y) = \exp(x + y)$

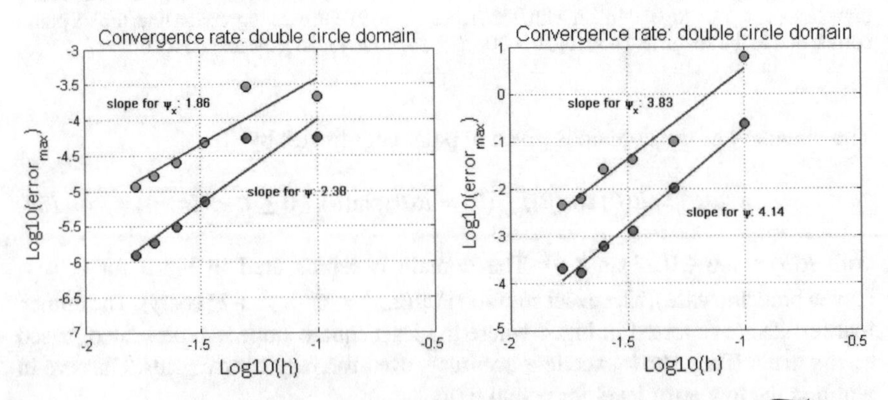

Fig. 6 Double circle shaped domain: linear regression and convergence rate for $\|\Psi - \widetilde{(\psi_{ex})}\|_\infty$ and $\|\Psi_x - \widetilde{(\psi_{x,ex})}\|_\infty$ with • *Left* $\psi_{ex}(x, y) = \exp(x + y)$ • *Right* $\psi_{ex}(x, y) = 10(x^5 \sin(4\pi y) + \frac{y^4}{1+x^2})$. For each regression line, the six points correspond to the six grids 10×10, 20×20, 30×30, 40×40, 50×50 and 60×60

References

1. Ben-Artzi, M., Chorev, I., Croisille, J.-P., Fishelov, D.: A compact difference scheme for the biharmonic equation in planar irregular domains. SIAM J. Numer. Anal. **47**, 3087–3108 (2009)
2. Ben-Artzi, M., Croisille, J.-P., Fishelov, D.: A fast direct solver for the biharmonic problem in a rectangular grid. SIAM J. Sci. Comput. **31**, 303–333 (2008)
3. Ben-Artzi M., Croisille J.-P., Fishelov, D.: Navier-Stokes Equations in Planar Domains. Imperial College Press (2013)
4. Chen, G., Li, Z., Lin, P.: A fast finite difference method for biharmonic equations on irregular domains and its application to an incompressible Stokes flow. Adv. Comput. Math. **2008**, 113–133 (2008)

Ant Colony Optimization Algorithm for 1D Cutting Stock Problem

Georgi Evtimov and Stefka Fidanova

Abstract Every day different companies in industry have to solve many optimization problems. One of them is cutting out of linear materials, like steel or aluminum profiles, steel or wood beams and so on. It is so called cutting stocks problem (CSP). It is well known NP-hard combinatorial optimization problem. The accurate and fast cutting out is very important element from the working process. The aim in CSP is to cut items from stocks of certain length, minimizing the total number of stocks (waste). The computational time increases exponentially when the number of items increase. Finding the optimal solution for large-sized problems for a reasonable time is impossible. Therefore, exact algorithms and traditional numerical methods can be apply of only on very small problems. Mostly appropriate methods for this kind of problems are methods based on stochastic search or so called metaheuristic methods. We propose a variant of Ant Colony Optimization (ACO) algorithm to solve linear cutting stocks problem.

1 Introduction

The 1D cutting stocks problem (CSP) is an important industrial problem. It appears in paper industry with cutting paper roles, in building construction cutting steel bars and cables. More precisely it is multiple stocks size cutting stocks problem. The aim is to reduce the waste and thus to minimize the expense of the producers, by cutting plan considering that the objects have different size.

Various methods have been proposed to solve this problem, from exact methods to metaheuristics. As an exact algorithms are applied linear programming [7] and branch and bound [10]. These methods can be applied only for small problems. When the number of profiles and bars increases the calculation time increases very fast,

G. Evtimov · S. Fidanova (✉)
Institute of Information and Communication Technologies,
Bulgarian Academy of Sciences, Acad. G. Bonchev Str., bl. 25A, 1113 Sofia, Bulgaria
e-mail: stefka@parallel.bas.bg

G. Evtimov
e-mail: gevtimov@abv.bg

K. Georgiev et al. (eds.), *Advanced Computing in Industrial Mathematics*, Studies in Computational Intelligence 728,
https://doi.org/10.1007/978-3-319-65530-7_3

therefore the exact methods are impractical for industrial use. So the scientists try to apply metaheuristic methods, which find close to optimal solutions using reasonable computational resources like time and memory. A number of metaheuristics have been applied on CSP. Some authors apply evolutionary algorithms including Genetic algorithm [2, 5, 8, 12], others apply Tabu search and Simulated annealing [6]. On this work we will propose a variant of Ant Colony Optimization (ACO) algorithm to solve CSP.

The first ACO algorithm was developed by Dorigo in his PhD thesis in 1992. During the years it has been improved several times and were created different variants of ACO algorithm. ACO was successfully applied on various combinatorial optimization problems. It is constructive method which does not need initial solution. ACO is very competitive method and outperforms other methods, especially when it is applied on combinatorial optimization problems with strong constraints [1].

The rest of the paper is organized as follows. In Sect. 2 the CSP problem is formulated. In Sect. 3 we describe the ACO algorithm and its application on CSP. In Sect. 4 we show experimental results and comparison with other algorithms is done. In Sect. 5 is a conclusion and some directions for future work are proposed.

2 Problem Formulation

One dimensional CSP (1D-CSP) has many applications in industry. The problem is NP-complete [4]. In this problem all used stocks length must be cut as much as possible. The remaining is as cutting waste and need to be minimized. Reduction of the cutting waste is a main goal of the 1D-CSP. Let there are n demands and m stocks. The stocks length is $d_j, j = 1, \dots, m$, s_i, $i = 1, \dots, n$, is a order length, n_i is a required number of orders, with length s_i. The 1D CSP can be defined as follows:

$$\min \sum_{i=1}^{n} \sum_{j=1}^{m} (d_j - x_{ij} s_i) \tag{1}$$

$$\sum_{j=1}^{m} x_{ij} = n_i, \; i = 1, \dots, n \tag{2}$$

where x_{ij} is the number of orders with a length s_i, that are cut from the stocks j.

Minimization of the waste is the objective function of the problem. The constraint guarantee the cutting needs to be satisfied. When the all stocks have the same length, than the cutting with minimal waste is equal to the cutting with minimal stocks.

Because the problem is NP-complete the computational time increase exponentially, and finding the optimal solution with some exact method or traditional numerical method is unpractical. Even finding feasible area of solutions is hard. Therefore on this type of problems normally are applied metaheuristic methods to compute near optimal solutions for a reasonable time. We propose a variant of Ant Colony Optimization algorithm to solve 1D cutting stocks problem.

3 Ant Colony Optimization

ACO is one of the most used and most successful metaheuristics [1]. It is applied on various combinatorial optimization problems coming from real life and industry. Examples of optimization problems are Traveling Salesman Problem [11], Vehicle Routing [13], Minimum Spanning Tree [9], Multiple Knapsack Problem [3], etc.

ACO is nature inspired methodology, which uses ideas from real ants behavior. When the ants look for a food they deposit a chemical substance, called pheromone, on their way back. After, they follow the path with stronger pheromone concentration. Thus the ants can find the shorter path between the nest and the food.

ACO represents a team of intelligent agents, which simulate ants behavior. The problem is represented by graph and the agents walk around it to solve the problem, using mechanisms of cooperation and adaptation. ACO is constructive method and it does not need initial solution. It is very appropriate for problems with strong constraints. The algorithm is iterative. Every ant constructs its solution starting from random position in a graph of the problem. After, it applies probabilistic rule, called transition probability to include next nodes in the solution till the solution is completed. At the end of every iteration the pheromone quantity is updated. The structure of the ACO algorithm is shown by the pseudo-code below (Fig. 1).

The transition probability $p_{i,j}$, to choose the node j when the current node is i, is based on the heuristic information $\eta_{i,j}$ and the pheromone trail level $\tau_{i,j}$ of the move, where $i, j = 1, \ldots, n$. The heuristic information represents the a priory knowledge of the problem and the pheromone corresponds to the ants experience from previous iterations to solve the problem:

Fig. 1 Pseudocode for ACO

```
Ant Colony Optimization
Initialize number of ants;
Initialize the ACO parameters;
while not end-condition do
      for k=0 to number of ants
          ant k choses start node;
          while solution is not constructed do
                ant k selects higher probability node;
          end while
      end for
      Update-pheromone-trails;
end while
```

$$p_{i,j} = \frac{\tau_{i,j}^{a}\eta_{i,j}^{b}}{\sum_{k \in allowed} \tau_{i,k}^{a}\eta_{i,k}^{b}}, \tag{3}$$

The higher the value of the pheromone and the heuristic information, the more profitable it is to select this move and resume the search. In the beginning, the initial pheromone level is set to a small positive constant value τ_0; later, the ants update this value after completing the construction stage. ACO algorithms adopt different criteria to update the pheromone level.

The pheromone trail update rule is given by:

$$\tau_{i,j} \leftarrow \rho\tau_{i,j} + \Delta\tau_{i,j}, \tag{4}$$

where ρ models evaporation in the nature and $\Delta\tau_{i,j}$ is a new added pheromone which is proportional to the quality of the solution. First the quantity of the pheromone is decreased to decrease the influence of the old information (history of the ants to construct solutions). After is added new pheromone, which intensify the search around the good so far solutions.

In our implementation on every iteration every ant choses a stock and an order in a random way. After is applied transition probability rule till the rest from the stock is shorter than the shortest order (no more possibility to cut), it is the waste. After that the ant chooses in a random way the next stock and order and continues applying the transition probability for the next cuts on the same stock. The ant does this, till no more orders are. The orders corresponds to the nodes of the graph of the problem and the arcs fully connect the nodes. We deposit the pheromone on the arcs, to show the sequence of the cutting orders. At the end of every iteration we update the pheromone. The new added pheromone is inversely proportional to the quantity of the waste. Thus the elements of the solution with less waste will receive more pheromone than others and will be more desirable in the next iteration. The random choose of the stocks and the first cutted order on every stock is a kind of diversification of the search in a search space. The pheromone updating is a kind of intensification of the search around the best so far solution. The heuristic information, which we apply in our application is equal to the length of the order. Thus the algorithm prefer the longest possible order which can be cutted from the rest of the stock. On the first iteration the quantity of the pheromone is the same for all edges and the algorithm works in a greedy way. From the second iteration the algorithm start to take in to account the ants experience, because the pheromone is different after the updating.

4 Experimental Results

We test our algorithm on real data coming from steel structure, composed from steel profiles and steel plates. The test example of the task are the profiles from the structure (Fig. 2).

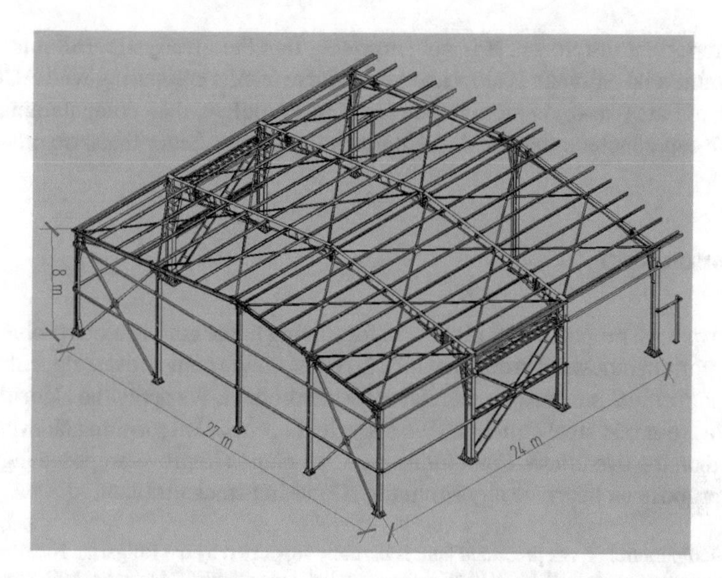

Fig. 2 Steel Structure, composed from steel profiles and steel plates

Table 1 ACO algorithm parameters

Parameter	Value
Number of ants	2
Initial pheromone	0.5
Number of iterations	100
Evaporation	0.1

The profiles are 924 pieces and 37 types. The orders to cut are from 2 to 672 pieces from the profiles of different types (length), from 50 to 6443 mm. The stocks are 12 m bars. In this case the minimum waste is equal to find a solution with minimal used bars. The algorithm parameters of our application are shown on Table 1. The algorithm is run on desktop computer with 2.8 GHz CPU.

We run 30 times our ACO algorithm and greedy algorithm and compare them with results achieved by one commercial software used by professionals. Regarding the Table 2 we observe that ACO algorithm achieves better result than the two other algorithms. We report the number of used bars. The ACO achieves solution with two bars less, but for producers one bar is more than thousand dollars. Thus even one

Table 2 Results comparison

Properties	Greedy	Commercial	ACO
Time in minutes	2	20	10
Used bars	234	234	232

bar is important difference. It is not surprising that the greedy algorithm is fastest. The commercial software is slow and gives worse results comparing with ACO. The number of bars (waste) is more important in this problem, than computational time. Thus we can conclude that the ACO algorithm performs better than two others.

5 Conclusion

In this paper we propose ACO algorithm for solving linear cutting stock problem. It is very important industrial problem which is NP-hard. We compare our algorithm with greedy algorithm and one commercial software product. We apply the algorithms on real data from real steel structure. We show that our ACO algorithm achieves better results than the two others. For a future work we plan to improve proposed algorithm and to propose an heuristic algorithm for 2D cutting stock problem.

Acknowledgements Work presented here is partially supported by the Bulgarian National Scientific Fund under Grants DFNI I02/20 "Efficient Parallel Algorithms for Large Scale Computational Problems" and DN 02/10 "New Instruments for Data Mining and their Modeling".

References

1. Dorigo, M., Maniezzo, M., Colorni, A.: The ant system: optimization by a colony of cooperating agents. IEEE Trans. Syst. Man Cybern. B **26**(1), 29–41 (1996)
2. Falkenauer, E.: A hybrid grouping genetic algorithm for bin packing. J. Heuristics **2**, 5–30 (1996)
3. Fidanova, S.: Evolutionary algorithm for multiple Knapsack problem. In: International Conference Parallel Problems Solving from Nature, Real World Optimization Using Evolutionary Computing, Granada, Spain (2002). ISBN No 0-9543481-0-9
4. Gradisar, M., Resinovic, G., Kljajic, M.: Evaluation of algorithms for one-dimensional cutting. Comput. Oper. Res. **29**(9), 1207–1220 (2002)
5. Hinterding, R., Khan, L.: Genetic algorithms for cutting stock problems: with and without contiguity. In: Yao, X. (ed.) Progress in Evolutionary Computation, pp. 166–186. Springer, Berlin, Germany (1995)
6. Jahromi, M.H., Tavakkoli-Moghaddam, R., Makui, A., Shamsi, A.: Solving an one-dimensional cutting stock problem by simulated annealing and Tabu search. J. Ind. Eng. Int. **8**(1), paper 24 (2012)
7. Kantorovicc, L.: Mathematical methods of organizing and planing production. Manag. Sci. **6**, 366–422 (1960)
8. Reeves, C.: Hybrid genetic algorithms for bin-packing and related problems. Ann. Oper. Res. **63**, 371–396 (1996)
9. Reiman, M., Laumanns, M.: A hybrid ACO algorithm for the capacitate minimum spanning tree problem. In: Proceedings of First International Workshop on Hybrid Metaheuristics, pp. 1–10, Valencia, Spain (2004)
10. Scheithauer, G., Terno, J.: A branch&bound algorithm for solving one-dimensional cutting stock problem exactly. Appl. Math. **23**(2), 151–167 (1995)

11. Stutzle, T., Dorigo, M.: ACO algorithm for the traveling salesman problem. In: Miettinen, K., Makela, M., Neittaanmaki, P., Periaux, J. (eds.) Evolutionary Algorithms in Engineering and Computer Science, pp. 163–183. Wiley (1999)
12. Vink, M.: Solving combinatorial problems using evolutionary algorithms (1997). http://citeseer.nj.nec.com/vink97solving.html
13. Zhang, T., Wang, S., Tian, W., Zhang, Y.: ACO-VRPTWRV: a new algorithm for the vehicle routing problems with time windows and re-used vehicles based on ant colony optimization. In: Sixth International Conference on Intelligent Systems Design and Applications, pp. 390–395. IEEE Press (2006)

2D Optimal Cutting Problem

Georgi Evtimov and Stefka Fidanova

Abstract Good management of industrial processes lead to optimization problems. Some of them are NP-hard and needs special algorithms to be solved. One such problem is cutting stock problem (CSP). The accurate and fast cutting out with less possible waste is very important element from the working process. The aim is to cut 2D items from rectangular stock, minimizing the waste. The problem is very difficult and the most of the authors solve the simplified version of the problem when the items are rectangular. The computational time increases exponentially when the number of items increase. Finding the optimal solution for large-sized problems for a reasonable time is impossible. Therefore exact algorithms and traditional numerical methods can be apply only on very small problems, less than 100 items. We propose an approximate algorithm which solve the problem when the items are polygons.

1 Introduction

The 2D CSP is an important industrial problem. Most popular is 2D CSP where the stocks and the items are rectangular. This problem appear in paper industry and in glass industry [6], container loading, Very Large Scale Integration (VLSI) design, and various scheduling tasks [7]. The problem becomes more complicate when the items are not rectangular. They can be any polygon, convex or concave. This problem arises in clothes production, plates in building constructions, shoes production and so on. In some applications the rotation is not possible, while in other it is possible and can be used for minimizing the waste.

G. Evtimov (✉) · S. Fidanova
Institute of Information and Communication Technologies,
Bulgarian Academy of Sciences, Acad. G. Bonchev Str., bl. 25A, 1113 Sofia, Bulgaria
e-mail: gevtimov@abv.bg

S. Fidanova
e-mail: stefka@parallel.bas.bg

K. Georgiev et al. (eds.), *Advanced Computing in Industrial Mathematics*, Studies in Computational Intelligence 728,
https://doi.org/10.1007/978-3-319-65530-7_4

In [6] the main topic is a two-dimensional orthogonal packing problem, where a fixed group of small rectangles must be fitted into a large rectangle so that, most of the material is used, and the unused area of the large rectangle is minimized. The algorithm combines a replacement method with a genetic algorithm. In [1] a number of heuristic algorithms for two-dimensional cutting problems (on large scales) are developed. In this study, there is a large primary stock that has to be cut into smaller pieces, so as to maximize the number of the pieces. They developed a greedy randomized adaptive search procedure. Cintra et al. [3] propose an exact algorithm based on dynamic programming. This kind of algorithms are appropriate for small problems, because the problem is NP-hard. For these problems are more appropriate to apply some method based on stochastic search. Stochastic search do not guarantee finding optimal solution, but they find quickly acceptable for the practitioners solution. Dusberger and Raidl [4, 5] propose two metaheuristic algorithms based on variable neighborhood search.

All this mentioned algorithms solve the problem, when the items are rectangular. In this paper we propose an algorithm, which is more realistic. It finds a solution when the items are polygons and their shape may be different from rectangular. We test our algorithm on real data. Our algorithm is compared with one commercial software and show that ours finds better solution.

The rest of the paper is organized as follows. In Sect. 2 the CSP problem is formulated. In Sect. 3 we describe our algorithm for solving CSP. In Sect. 4 we show experimental results and comparison with other algorithms are done. Section 5 is a conclusion and some directions for future work are proposed.

2 Problem Formulation

Most of the authors solve simplified variant of the cutting stock problem where the items are rectangular. This problem arise in paper and glass industries. Others complete the ordered items to rectangles, but it is not effective [2]. When some shape is completed to rectangular, the surface of the received rectangular can be more than two times larger than the surface of the initial item. In some of commercial products is possible after completing the large half of the items to rectangular, to verify if some of the smaller items can be included in some of the rectangular without crossing. It improves the algorithm, but it continue to be effective.

In this paper we expect that is given rectangular sheet with fixed width and infinite length. The set $E = \{i_1, i_2, \dots, i_n\}$ of ordered items are polygons, which can be convex and concave. Examples of ordered items are shown on Fig. 1. The left bottom corner of the sheet has coordinates $(0, 0)$.

The elements are specified by their nodes and demands d_i, for $i = 1, \dots, n$. The elements can be rotated.

The objective is to find a cutting pattern P, the arrangement of the elements in E on the stock sheet, without overlap with a minimal waste. Let the width of the sheet is fixed to be x and the cutting height in P to be y. The area of ordered items is fixed,

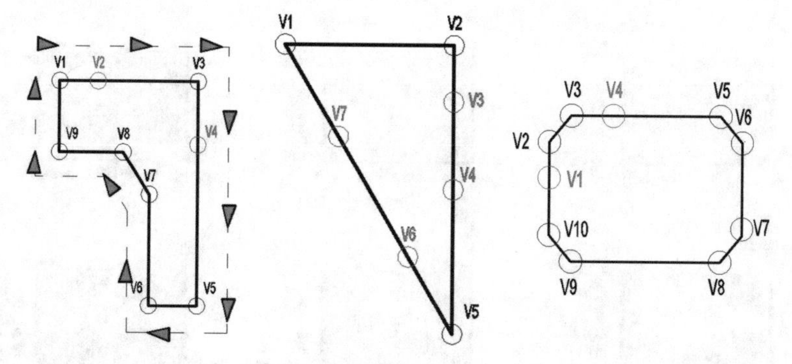

Fig. 1 The polygons which will be cut

thus the cutting pattern with a minimal waste is a cutting pattern, with a minimal cutting height. So the objective function $C(P)$ is:

$$C(P) = min(y). \tag{1}$$

The solution, cutting pattern P, is represented by cutting sequence and coordinates of the nodes of the cutting items.

3 Proposed Algorithm

The two dimensional cutting stock problem is an important optimization problem which arises in many industries. Even the simplified problem, where the cutting items are rectangular is NP-hard [2, 8]. In this paper we propose an algorithm which finds solutions, when the cutting items are polygons. At the beginning our algorithm verifies for every polygon:

- if the polygon is self-crossing;
- if the polygon is a line;
- are there redundant points.

The cutting items is not necessary to be convex, it must be non self-crossing. Each edge is linear, and the polygon is described by its vertexes. The sheet, from which we will cut the item is rectangular with fixed width and infinity length. The algorithm chooses a random item from the set E of items and puts one of its nodes on the point with minimal high (y) in the sheet. Let the number of the nodes of the chosen item are m. We will translate the item m times and will rotate it as it is shown on Fig. 2. The number of combinations without mirroring is m^2. The algorithm accept the positioning where all points inside the polygon are in the sheet of cutting and where the cutting hight (y) is minimal. Regarding Fig. 2 in seven of the cases the cutting item is outside the stock sheet. Thus these positioning are not eligible.

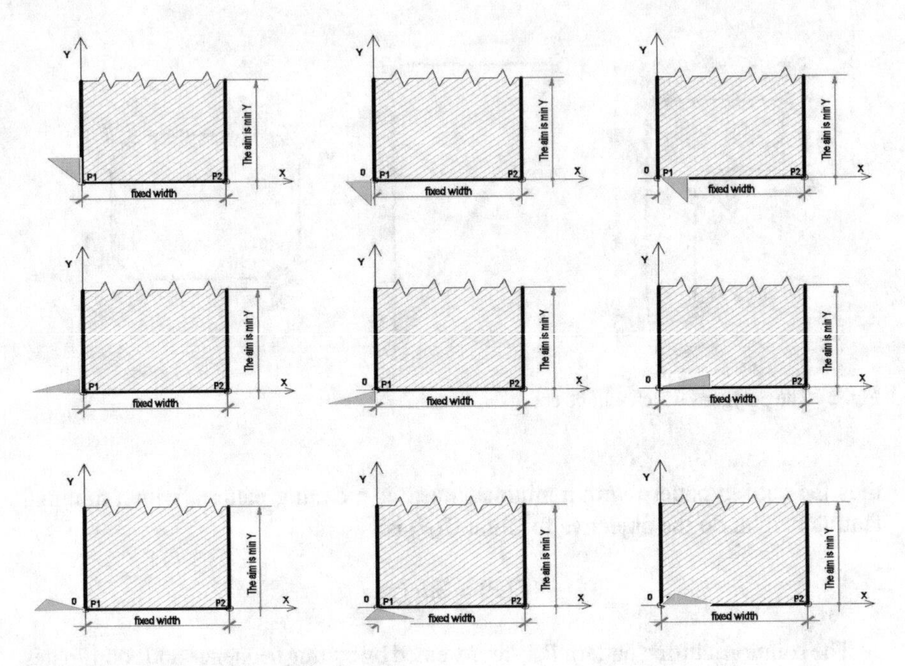

Fig. 2 Translations and rotations of a polygon

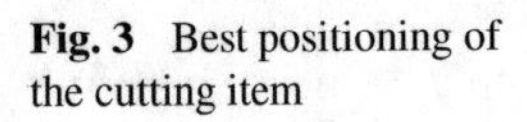

Fig. 3 Best positioning of the cutting item

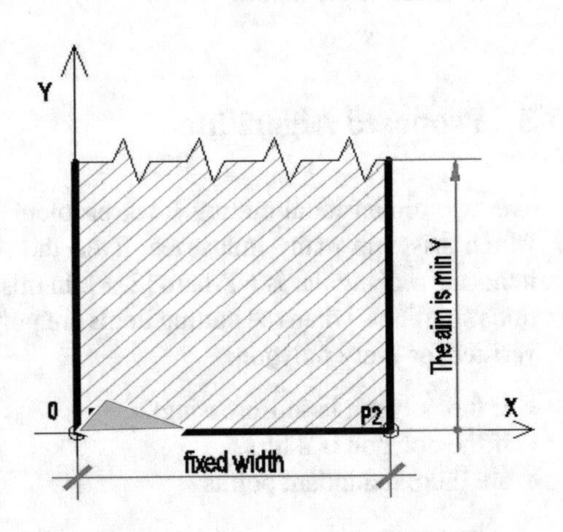

There are two acceptable positioning and the algorithm chose the one with smaller cutting hight (y).

In our example the cutting with minimal y is shown on Fig. 3.

On the next steps we again chose in a random way a polygon from the set E of items and try to fix it on the point (node) with a minimal y from the stock sheet. If there are more than one points with minimal y, the algorithm chooses one of them in a random way. If it is impossible to position chosen item on this point we chose the next one with minimal y. We do this till no more items in the set E.

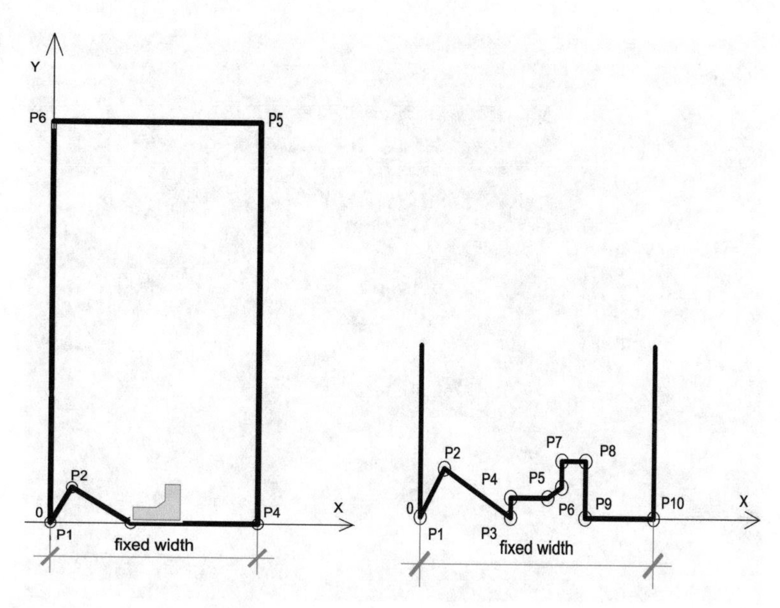

Fig. 4 Algorithm performance

Figure 4 illustrates the algorithm performance. Proposed algorithm can be applied when the bottom part of the stock sheet is not a straight line. This situation arise when cutting of some order is finished and later the producer is prepared to cut a new order. With vary small changes our algorithms can solve the variant of the problem, where there are several stock sheets with fixed length.

4 Experimental Results

We test our algorithm on real data coming from steel structure, composed from steel profiles and steel plates, Fig. 5. The test example of the task are the plates from the structure. Some of them are convex and others are concave polygons. The plates are 1958 pieces and 242 types. The overall area of the plates is 129,053,789 mm^2. We cut the plates from steel sheet with fixed width equal to 1500 mm. In this case the minimum waste is equal to find a solution with minimal y and respectively minimal filling factor. The filling factor is the ratio between the sum of the area of the all plates and the cutting area ($x \times y$). We take in to account that the cut width is 5 mm. Before running the optimization procedure we process the input data. The input data are polygons, described by points. Very often a plate is described with more points than is the number of their nodes. We verify the input data and we remove needless points. At the end the plates are described only with their nodes. The algorithm is run on desktop computer with 2.8 GHz CPU.

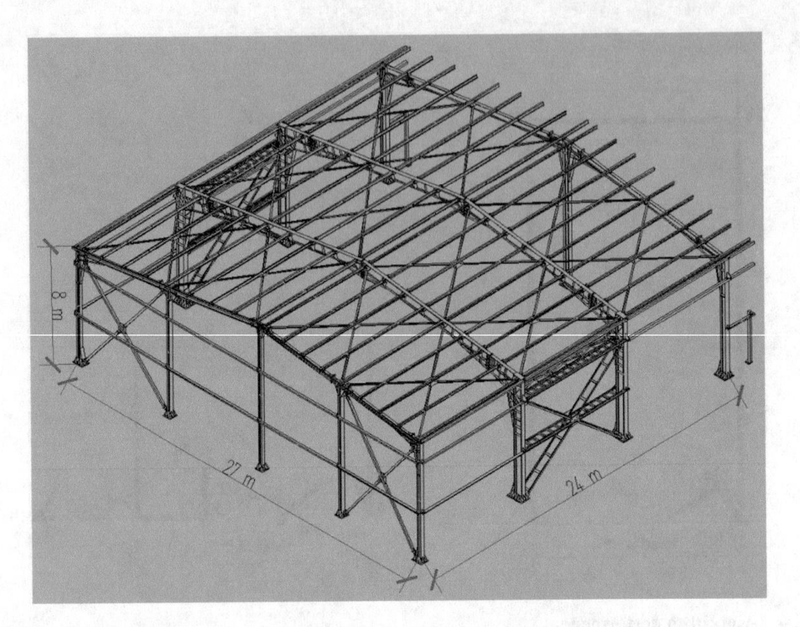

Fig. 5 Steel structure, composed of steel profiles and steel plates

Table 1 Results comparison

Properties	Commercial algorithm	Proposed algorithm
Cutted hight	200,014 mm	170,273 mm
Filling factor	0.43	0.505

We run the proposed algorithm 30 times and we compare achieved best results with results achieved by one commercial software used by professionals. We chose a professional software, which do not complete the items to rectangular. We report the value of cutted hight (y) and filling factor. Regarding the Table 1 we observe that proposed algorithm achieves better result than the commercial one. Our algorithm achieves solution with smaller hight and respectively less filling factor, which is equal to solution with less waste. We can conclude that our algorithm receives very encouraging results.

5 Conclusion

In this paper we propose an algorithm for solving 2D cutting stock problem. It is very important industrial problem which is NP-hard. The cutting items are convex and concave polygons. We compare our algorithm with one commercial software product. We apply the algorithms on real data from real steel structure. We show that our algorithm achieves better results than the other. For a future work we plane to improve proposed algorithm and to propose an heuristic algorithm for 2D cutting stock problem.

Acknowledgements Work presented here is partially supported by the Bulgarian National Scientific Fund under Grants DFNI I02/20 "Efficient Parallel Algorithms for Large Scale Computational Problems" and DFNI DN 02/10 "New Instruments for Data Mining and their Modeling".

References

1. Alvarez-Valdes, R., Parajon, A., Tamarit, J.M.: A computational study of heuristic algorithms for two-dimensional cutting stock problems. In: 4th Metaheuristics International Conference (MIC2001), pp. 16–20 (2001)
2. Alvarez-Valdes, R., Parreno, F., Tamarit, J.M.: A Tabu search algorithm for two dimensional non-guillotine cutting problems. Eur. J. Oper. Res. **183**(3), 1167–1182 (2007)
3. Cintra, G., Miyazawa, F., Wakabayashi, Y., Xavier, E.: Algorithms for two-dimensional cutting stock and strip packing problems using dynamic programming and column generation. Eur. J. Oper. Res. **191**, 61–85 (2008)
4. Dusberger, F., Raidl, G.R.: A variable neighborhood search using very large neighborhood structures for the 3-staged 2-dimensional cutting stock problem. In: Hybrid Metaheuristics. Lecture Notes in Computer Science, vol. 8457, pp. 85–99. Springer (2014)
5. Dusberger, F., Raidl, G.R.: Solving the 3-staged 2-dimensional cutting stock problem by dynamic programming and variable neighborhood search. Electron. Notes Discret. Math. **47**, 133–140 (2015)
6. Gonalves, J.F.: A hybrid genetic algorithm-heuristic for a two-dimensional orthogonal packing problem. Eur. J. Oper. Res. **183**(3), 1212–1229 (2007)
7. Lodi, A., Martello, S., Vigo, D.: Recent advances on two-dimensional bin packing problems. Discret. Appl. Math. **123**, 379–396 (2002)
8. Parmar, K., Prajapati, H., Dabhi, V.: Cutting stock problem: a survey of evolutionary computing based solution. In: Proceedings of Green Computing Communication and Electrical Engineering. (2014). doi:10.1109/ICGCCEE.2014.6921411

Error Estimate in the Iterative Numerical Method for Two-Dimensional Nonlinear Hammerstein-Fredholm Fuzzy Functional Integral Equations

Atanaska Georgieva, Albena Pavlova and Iva Naydenova

Abstract In this paper, we prove the convergence of the method of successive approximations used to approximate the solution of two-dimensional nonlinear Hammerstein-Fredholm fuzzy functional integral equations. We present an iterative procedure based on quadrature rectangles to solve such equations. The error estimation of the proposed method is given in terms of uniform and partial modulus of continuity. Finally, an illustrative numerical experiment confirms the theoretical results and demonstrates the accuracy of the method.

1 Introduction

The concepts of fuzzy integral and differential equations have been studied by many mathematicians and authors. The study of fuzzy integral equations begins with the investigations performed by of Kaleva [9], Seikkala [13], Goetshel and Voxman [8] and others. The interest in fuzzy Fredholm integral equations is based primarily on its applications in fuzzy financial and economic systems [4]. Banach's fixed point theorem and method of successive approximation are applied in the problem of the existence and uniqueness of the solution(see [2, 6, 10]).

The numerical methods for solving fuzzy integral equations involve various techniques. The method of successive approximations and other iterative techniques are

A. Georgieva (✉) · I. Naydenova
FMI, University of Plovdiv Paisii Hilendarski, 24 Tzar Asen, 4003 Plovdiv, Bulgaria
e-mail: atanaska@uni-plovdiv.bg

I. Naydenova
e-mail: iva.naydenova@asm32.info

A. Pavlova
Department of MPC Technical University-Sofia, Plovdiv Branch, 25 Tzanko Djustabanov Str., 4000 Plovdiv, Bulgaria
e-mail: akosseva@gmail.com

K. Georgiev et al. (eds.), *Advanced Computing in Industrial Mathematics*, Studies in Computational Intelligence 728,
https://doi.org/10.1007/978-3-319-65530-7_5

applied in [6, 7]. Since many problems in engineering and applied sciences can be put in the form of two-dimensional fuzzy integral equations, it is important to develop numerical methods for solving such integral equations. In this paper, we investigate the two-dimensional nonlinear Hammerstein-Fredholm fuzzy functional integral equation

$$\begin{aligned} F(s,t) &= g(s,t) \oplus f(s,t,F(s,t)) \oplus \\ &\oplus (FR)\int_c^d (FR)\int_a^b K(s,t,x,y) \odot H(x,y,F(x,y))dxdy, \end{aligned} \tag{1}$$

where $K(s,t,x,y)$ is an arbitrary positive kernel on $[a,b] \times [c,d] \times [a,b] \times [c,d]$, $g : [a,b] \times [c,d] \to \mathbf{R}_{\mathscr{F}}, f,\ H : [a,b] \times [c,d] \times \mathbf{R}_{\mathscr{F}} \to \mathbf{R}_{\mathscr{F}}$ are continuous fuzzy-number valued functions.

The existence and uniqueness of the solution is proven by Banach's fixed point theorem. We approximate the solution of the equation using the quadrature formula of rectangles and the method of successive approximations. The error estimation of the iterative method is obtained in terms of uniform and partial modulus of continuity, proving the convergence of the method. The error estimate obtained in this paper is expressed in terms of the modulus of continuity for g and K. We illustrate the iterative method of numerical experiment by testing the convergence and the numerical stability with respect to the choice of the first iterations. A numerical example is included in order to confirm the theoretical results of the test.

2 Preliminaries

Firstly, we present some notions and results about fuzzy numbers and fuzzy-number-valued functions.

Definition 1 [5, 8] A fuzzy number is a function $u : R \to [0,1]$ satisfying the following properties:

1. u is upper semi-continuous on R,
2. $u(x) = 0$ outside of some interval $[c,d]$,
3. there are the real numbers a and b with $c \le a \le b \le d$, such that u is increasing on $[c,a]$, decreasing on $[b,d]$, and $u(x) = 1$ for each $x \in [a,b]$,
4. u is fuzzy convex set (that is $u(\lambda x + (1-\lambda)y) \ge \min\{u(x), u(y)\}$, for all $x, y \in R$, $\lambda \in [0,1]$) and possess compact support $[u]^0 = \overline{\{x \in R : u(x) > 0\}}$, where $\overline{A}$ denotes the closure of A.

The set of all fuzzy numbers is denoted by $\mathbf{R}_{\mathscr{F}}$. Any real number $a \in R$ can be interpreted as a fuzzy number $\tilde{a} = \chi(a)$ and therefore $R \subset \mathbf{R}_{\mathscr{F}}$. The neutral element with respect to $\oplus$ in $\mathbf{R}_{\mathscr{F}}$ is denoted by $\tilde{0} = \chi_{\{0\}}$. For any $0 < r \le 1$ we point out the r-level set $[u]^r = \{x \in R : u(x) \ge r\}$,that is a closed interval, and $[u]^r = [u^r_-, u^r_+]$ for any $r \in [0,1]$. These r-level sets lead to the usual LU representation of a fuzzy number: $[u]^r = [u^r_-, u^r_+]$ for any $r \in [0, 1]$, where u_-, u_+ can be consider as functions $u_-, u_+ : [0, 1] \to \mathbf{R}$, such that u_- is increasing and u_+ is decreasing.

For $u, v \in \mathbf{R}_{\mathcal{F}}$, $k \in R$, the addition and the scalar multiplication are defined by $[u \oplus v]^r = [u]^r + [v]^r$, and $[k \odot u]^r = k.[u]^r$ for all $r \in [0, 1]$. In [1, 13] are given algebraic properties for any $u, v, w \in \mathbf{R}_{\mathcal{F}}$.

Definition 2 [1] For arbitrary fuzzy numbers $u = (u^r_-, u^r_+)$ and $v = (v^r_-, v^r_+)$ the quantity $D(u, v) = \sup_{r\in[0,1]} \max\{|u^r_- - v^r_-|, |u^r_+ - v^r_+|\}$ is the distance between u and v.

Theorem 1 *[14] The following properties of the above distance hold:*

1. *$(\mathbf{R}_{\mathcal{F}}, D)$ is a complete metric space,*
2. *$D(u \oplus w, v \oplus w) = D(u, v)$, for all $u, v, w \in \mathbf{R}_{\mathcal{F}}$,*
3. *$D(k \odot u, k \odot v) = |k|D(u, v)$, for all $u, v \in \mathbf{R}_{\mathcal{F}}$, for all $k \in R$,*
4. *$D(u \oplus v, w \oplus e) = D(u, w) + D(v, e)$, for all $u, v, w, e \in \mathbf{R}_{\mathcal{F}}$,*
5. *$D(u \oplus v, \tilde{0}) \le D(u, \tilde{0}) + D(v, \tilde{0})$, for all $u, v \in \mathbf{R}_{\mathcal{F}}$,*
6. *$D(k_1 \odot u, k_2 \odot u) = |k_1 - k_2|D(u, \tilde{0})$, for all $k_1, k_2 \in R$ with $k_1 k_2 \ge 0$ and $u \in \mathbf{R}_{\mathcal{F}}$.*

Guided by the property 5 from Theorem 1, in [3] Bede and Gal it is defined a function $\|.\|_{\mathcal{F}} : \mathbf{R}_{\mathcal{F}} \to R$ by $\|u\|_{\mathcal{F}} = D(u, \tilde{0})$ that has the properties of usual norms:

1. $\|u\|_{\mathcal{F}} \ge 0$, for all $u \in \mathbf{R}_{\mathcal{F}}$ and $\|u\|_{\mathcal{F}} = 0$ iff $u = \tilde{0}$,
2. $\|\lambda \odot u\|_{\mathcal{F}} = |\lambda| \odot \|u\|_{\mathcal{F}}$ and $\|u \oplus v\|_{\mathcal{F}} \le \|u\|_{\mathcal{F}} + \|v\|_{\mathcal{F}}$ for all $u, v \in \mathbf{R}_{\mathcal{F}}$, for all $\lambda \in R$,
3. $|\ \|u\|_{\mathcal{F}} - \|v\|_{\mathcal{F}}\ | \le D(u, v)$ and $D(u, v) \le \|u\|_{\mathcal{F}} + \|v\|_{\mathcal{F}}$, for all $u, v \in \mathbf{R}_{\mathcal{F}}$.

For any fuzzy-number-valued function $f : A = [a, b] \times [c, d] \to \mathbf{R}_{\mathcal{F}}$ we can define the functions $\underline{f(.,.,r)}, \overline{f(.,.,r)} : A \to R$, $r \in [0, 1]$ by $\underline{f(s, t, r)} = f(s, t, r)^r_-$, $\overline{f(s, t, r)} = f(s, t, r)^r_+$ for all $(s, t) \in A$, for all $r \in [0, 1]$. These functions are called the left and right r-level functions of f.

Definition 3 [12] A fuzzy-number-valued function $f : A \times [c, d] \to \mathbf{R}_{\mathcal{F}}$ is said to be continuous at $(s_0, t_0) \in A$ if for each $\varepsilon > 0$ there is $\delta > 0$ such that $D(f(s, t), f(s_0, t_0)) < \varepsilon$ whenever $\sqrt{(s - s_0)^2 + (t - t_0)^2} \le \delta$. If f be continuous for each $(s, t) \in A$, then we say that f is continuous on A.

On the set $C(A, \mathbf{R}_{\mathcal{F}}) = \{f : A \to \mathbf{R}_{\mathcal{F}} : f \text{ is continuous}\}$ it is defined the metric $D^*(f, g) = \sup_{(s,t)\in A} D(f(s, t), g(s, t))$, for all $f, g \in C(A, \mathbf{R}_{\mathcal{F}})$. This metric is called the uniform distance between fuzzy-number-valued functions and (X, D^*) is a complete metric space.

Definition 4 A fuzzy-number-valued function $f : A \to \mathbf{R}_{\mathcal{F}}$ is called bounded iff there is $M \ge 0$ such that $D(f(s, t), \tilde{0}) \le M$ for all $(s, t) \in A$.

Definition 5 [14] Let $f : A \to \mathbf{R}_{\mathcal{F}}$, for $\Delta_x^n : a = x_0 < x_1 < \cdots < x_n = b$ and Δ_y^n : $c = y_0 < y_1 < \cdots < y_n = d$, be two partitions of the intervals $[a, b]$ and $[c, d]$, respectively. Let one consider the intermediate points $\xi_i \in [x_{i-1}, x_i]$ and $\eta_j \in [y_{j-1}, y_j]$, $i = 1, ..., n; j = 1, ..., n$, and $\delta : [a, b] \to R_+$ and $\sigma : [c, d] \to R_+$. The divisions $P_x = ([x_{i-1}, x_i]; \xi_i), i = 1, ..., n$, and $P_y = ([y_{j-1}, y_i]; \eta_j), j = 1, ..., n$, denoted shortly by $P_x = (\Delta^n, \xi)$ and $P_y = (\Delta^n, \eta)$ are said to be δ-fine and σ-fine, respectively, if $[x_{i-1}, x_i] \subseteq (\xi_i - \delta(\xi_i), \xi_i + \delta(\xi_i))$ and $[y_{j-1}, y_j] \subseteq (\eta_j - \sigma(\eta_j), \eta_j + \sigma(\eta_j))$.

The function f is said to be two-dimensional Henstock integrable to $I \in \mathbf{R}_{\mathcal{F}}$ if for every $\varepsilon > 0$ there are functions $\delta : [a, b] \to \mathbf{R}_{\mathcal{F}}$ and $\sigma : [c, d] \to \mathbf{R}_{\mathcal{F}}$ such that for any δ-fine and σ-fine divisions we have $D(\sum_{j=1}^{n} \sum_{i=1}^{n} (x_i - x_{i-1})(y_j - y_{j-1}) \odot f(\xi_i, \eta_j), I) < \varepsilon$, where $\sum$ denotes the fuzzy summation. Then, I is called the two-dimensional Henstock integral of f and is denoted by $I(f) = (FH) \int_c^d (FH) \int_a^b f(s, t) dsdt$.

If the above δ and σ are constant functions, then one recaptures the concept of Riemann integral. In this case, $I \in \mathbf{R}_{\mathcal{F}}$ will be called two-dimensional integral of f on A and will be denoted by $(FR) \int_c^d (FR) \int_a^b f(s, t) dsdt$.

In [14], the authors introduced and concept of the Henstock integral for a fuzzy number-valued function.

Lemma 1 *[11] If $f : A \to \mathbf{R}_{\mathcal{F}}$ is a fuzzy-Henstock integrable bounded mapping then for any fixed $u \in [a, b]$ and $v \in [c, d]$ the function $\varphi_{u,v} : A \to R_+$ defined by $\varphi_{u,v}(s, t) = D(f(u, v), f(s, t))$ is Lebesgue integrable on A and* $D\left((FH) \int_c^d (FH) \int_a^b f(s, t) dsdt, (FH) \int_c^d (FH) \int_a^b g(s, t) dsdt \right) \leq (L) \int_c^d (L) \int_a^b D(f(s, t), g(s, t)) dsdt.$

Definition 6 [11] Let $f : A \to \mathbf{R}_{\mathcal{F}}$, be a bounded mapping, then the function $\omega_A(f, .) : R_+ \cup 0 \to R_+$ defined by $\omega_A(f, \delta) = \sup\{D(f(x, y), f(s, t)) : (x, y), (s, t) \in A; \sqrt{(x - s)^2 + (y - t)^2} \leq \delta\}$ is called the modulus of oscillation of f on A. In addition if $f \in C(A, \mathbf{R}_{\mathcal{F}})$, then $\omega_A(f, \delta)$ is called uniform modulus of continuity of f.

According to [11] the following properties hold

1. $D(f(x, y), f(s, t)) \leq \omega_A(f, \sqrt{(x - s)^2 + (y - t)^2})$ for any $(x, y), (s, t) \in A$,
2. $\omega_A(f, \delta)$ is a non-decreasing mapping in δ,
3. $\omega_A(f, 0) = 0$,
4. $\omega_A(f, \delta_1 + \delta_2) \leq \omega_A(f, \delta_1) + \omega_A(f, \delta_2)$ for any $\delta_1, \delta_2 \geq 0$,
5. $\omega_A(f, n\delta) \leq n\omega_A(f, \delta)$ for any $\delta \geq 0$ and $n \in N$,
6. $\omega_A(f, \lambda\delta) \leq (\lambda + 1)\omega_A(f, \delta)$ for any $\delta, \lambda \geq 0$,
7. If $[a_1, b_1] \times [c_1, d_1] \subseteq A$, then $\omega_{[a_1,b_1]\times[c_1,d_1]}(f, \delta) \leq \omega_A(f, \delta)$ for all $\delta \geq 0$.

Lemma 2 *[11] Let $f : A \to \mathbf{R}_{\mathcal{F}}$, be two-dimensional Henstock integrable, bounded mapping. Then the following inequality holds* $D((FR)\int_c^d (FR)\int_a^b f(s,t)dsdt, (b-a)(d-c) \odot f(\frac{a+b}{2}, \frac{c+d}{2})) \leq (b-a)(d-c)\omega_A(f, \frac{b-a}{2} \cdot \frac{d-c}{2})$.

Theorem 2 *[12] Let $f \in C(A \times A, \mathbf{R}_{\mathcal{F}}), g \in C(A, \mathbf{R}_{\mathcal{F}})$ and $h \in C(A, R_+)$, then the functions $h.g : A \to \mathbf{R}_{\mathcal{F}}$ and $P : A \to \mathbf{R}_{\mathcal{F}}$ given by $(h.g)(s,t) = h(s,t) \odot g(s,t)$, for all $(s,t) \in A$ and $P(s,t) = (FH)\int_c^d (FH)\int_a^b f(s,t,x,y)dxdy$ are continuous.*

3 Successive Approximations and the Iterative Algorithm

We denote by $X = C(A, \mathbf{R}_{\mathcal{F}}) = \{f : A \to \mathbf{R}_{\mathcal{F}} : f \text{ is continuous }\}$. The following conditions are imposed:

(i) $g \in C(A, \mathbf{R}_{\mathcal{F}}), f, H \in C(A \times \mathbf{R}_{\mathcal{F}}, \mathbf{R}_{\mathcal{F}})$ and $K \in C(A \times A, R_+)$,
(ii) there exists $\alpha_1 \geq 0$, such that $D(f(s,t,u), f(s,t,v)) \leq \alpha_1 D(u,v)$ for all $(s,t) \in A$,
(iii) there exists $\alpha_2 \geq 0$, such that $D(H(s,t,u), H(s,t,v)) \leq \alpha_2 D(u,v)$ for all $(s,t) \in A$,
(iv) $B = \alpha_1 + N_K \Delta \alpha_2 < 1$, where $\Delta = (b-a)(d-c)$ and $N_K \geq 0$ is such that $|K(s,t,x,y)| \leq N_K$, for all $(s,t), (x,y) \in A$, according to the continuity of K.

We define the operator $T : X \to X$ by

$$T(F)(s,t) = g(s,t) \oplus f(s,t,F(s,t)) \oplus \oplus (FR)\int_c^d (FR)\int_a^b K(s,t,x,y) \odot H(x,y,F(x,y))dxdy, \quad \text{for all } (s,t) \in A, \ F \in X. \tag{2}$$

Theorem 3 *Under the conditions (i)–(iv) the integral Eq. (1) has unique solution $F \in C(A, \mathbf{R}_{\mathcal{F}})$ and the sequence of successive approximations $\{F_m\}_{m \in N} \subset X$*

$$F_m(s,t) = g(s,t) \oplus f(s,t,F_{m-1}(s,t)) \oplus \oplus (FR)\int_c^d (FR)\int_a^b K(s,t,x,y) \odot H(x,y,F_{m-1}(x,y))dxdy \tag{3}$$

converges to F in X for any choice of $F_0 \in X$ and the following error estimates hold:

$$D(F(s,t), F_m(s,t)) \leq \frac{B^m}{1-B} D(F_1(s,t), F_0(s,t)), \text{ for all } (s,t) \in A, m \in N \tag{4}$$

$$D(F(s,t), F_m(s,t)) \leq \frac{B}{1-B} D(F_m(s,t), F_{m-1}(s,t)), \text{ for all } (s,t) \in A, m \in N. \tag{5}$$

If $F_0 = g$ then the estimate (4) becomes

$$D(F(s,t), F_m(s,t)) \leq \frac{B^m}{1-B}(B\|g\|_{\mathcal{F}} + \|f\|_{\mathcal{F}} + N_K \Delta \|H\|_{\mathcal{F}}), \tag{6}$$

where $\|f\|_{\mathcal{F}} = \sup_{(s,t)\in A} D(f(s,t,\tilde{0}),\tilde{0})$, $\|H\|_{\mathcal{F}} = \sup_{(s,t)\in A} D(H(s,t,\tilde{0}),\tilde{0})$.

Moreover, the sequence of successive approximations (3) is uniformly bounded and the solution F is bounded too.

Proof First, we prove that $T(X) \subset X$. For this purpose, let arbitrary $F \in X$, $(s_0, t_0) \in A$ and $\varepsilon > 0$. Since F is continuous and A is compact set, we infer that F is uniformly continuous, and according to the uniform continuity of H with respect to the first and second argument, it follows that there exists $\delta(\varepsilon) > 0$ such that for any $(s,t) \in A$ with $\sqrt{(s-s_0)^2 + (t-t_0)^2} < \delta(\varepsilon)$ we have $D(H(s,t,F(s,t)), H(s_0,t_0,F(s,t))) \leq \frac{\varepsilon}{2}$ and $D(F(s,t), F(s_0,t_0)) < \frac{\varepsilon}{2\alpha}$. Then, $D(H(s,t,F(s,t)), H(s_0,t_0,F(s_0,t_0))) \leq D(H(s,t,F(s,t)), H(s_0,t_0,F(s,t))) + D(H(s_0,t_0,F(s,t)), H(s_0,t_0,F(s_0,t_0))) \leq \frac{\varepsilon}{2} + \alpha_2$ $D(F(s,t), F(s_0,t_0)) \leq \frac{\varepsilon}{2} + \alpha_2 \frac{\varepsilon}{2\alpha_2} \leq \varepsilon$ and the function $U_F : A \to \mathbf{R}_{\mathcal{F}}$ defined by $U_F(s,t) = H(s,t,F(s,t))$ is continuous in (s_0,t_0). We infer that U_F is continuous on A for any $F \in X$. Analogously, the function $W_F : A \to \mathbf{R}_{\mathcal{F}}$ defined by $W_F(s,t) = f(s,t,F(s,t))$ is continuous on A for any $F \in X$. Applying Theorem 2 it follows that the function $K(s,t,.,.) \odot U_F(.,.) : A \to \mathbf{R}_{\mathcal{F}}$ is continuous on A for any $F \in X$. Using the same Theorem 2 we see that the function $V_F : A \to \mathbf{R}_{\mathcal{F}}$, defined by $V_F(s,t) = (FR)\int_c^d (FR)\int_a^b K(s,t,x,y) \odot U_F(x,y)dxdy$ is continuous on A for any $F \in X$. Since $g \in X$, we conclude that $T(F)$ is continuous on A for any $F \in X$. Now we prove that the operator T is a contraction. Let arbitrary $F, G \in X$. From Definition 3, condition 3 of Theorem 1, Lemma 1 and conditions (ii)-(iv) we have

$$D(T(F)(s,t), T(G)(s,t)) \leq D(g(s,t), g(s,t)) + D(f(s,t,F(s,t)), f(s,t,G(s,t))) +$$

$$+ D((FR)\int_c^d (FR)\int_a^b K(s,t,x,y) \odot H(x,y,F(x,y))dxdy,$$

$$(FR)\int_c^d (FR)\int_a^b K(s,t,x,y) \odot H(x,y,G(x,y))dxdy) \leq \alpha_1 D(F(s,t)), G(s,t)) +$$

$$+ \int_c^d \int_a^b |K(s,t,x,y)| D(H(x,y,F(x,y)), H(x,y,G(x,y)))dxdy \leq$$

$$\leq \alpha_1 D(F(s,t)), G(s,t)) + N_k \alpha_2 \int_c^d \int_a^b D(F(x,y), G(x,y))dxdy \leq$$

$$\leq \alpha_1 D^*(F,G) + N_K \Delta \alpha_2 D^*(F,G) \leq BD^*(F,G).$$

Therefore $D^*(T(F), T(G)) \leq BD^*(F,G)$ for all $F, G \in X$. Since $B < 1$, T is contraction. Applying the Banach's fixed point principle we obtain the existence of the uniqueness of the solution $F \in X$, of (1) and the uniform convergence of the sequence of successive approximations (3) to this solution in X, for any choice of the initial term of $F_0 \in X$. The inequality (4) and (5) we obtain from Banach's fixed point principle.

Choosing $F_0 = g$. For all $(s,t) \in A$ and the functions $f, H \in C(A \times \mathbf{R}_{\mathscr{F}}, \mathbf{R}_{\mathscr{F}})$, $g \in X$ we obtain
$D(f(s,t,g(s,t)),\tilde{0}) \leq D(f(s,t,g(s,t)),f(s,t,\tilde{0})) + D(f(s,t,\tilde{0}),\tilde{0}) \leq \alpha_1 \|g\|_{\mathscr{F}} + \|f\|_{\mathscr{F}}$ and $D(H(s,t,g(s,t)),\tilde{0}) \leq \alpha_2 \|g\|_{\mathscr{F}} + \|H\|_{\mathscr{F}}$. So,

$$D(F_1(s,t), F_0(s,t)) \leq$$

$$\leq D(f(s,t,g(s,t)),\tilde{0}) + D((FR)\int_c^d (FR)\int_a^b K(s,t,x,y) \odot H(x,y,g(x,y))dxdy, \tilde{0}) \leq$$

$$\leq B\|g\|_{\mathscr{F}} + \|f\|_{\mathscr{F}} + N_K \Delta \|H\|_{\mathscr{F}} i.e.$$

$$D(F_1(s,t), F_0(s,t)) \leq B\|g\|_{\mathscr{F}} + \|f\|_{\mathscr{F}} + N_K \Delta \|H\|_{\mathscr{F}}. \tag{7}$$

and condition (4) we obtain the inequality (6).

For arbitrary $(s,t) \in A$ and $m \in N$, we have $D(F_m(s,t), F_{m-1}(s,t)) \leq BD^*(F_{m-1}, F_{m-2})$ by induction we obtain

$$D(F_m(s,t), F_{m-1}(s,t)) \leq B^{m-1} D^*(F_1, F_0). \tag{8}$$

Then from (7) and (8) we obtain

$$\begin{aligned} D(F_m(s,t), F_0(s,t)) &\leq D(F_m(s,t), F_{m-1}(s,t)) + D(F_{m-1}(s,t), F_{m-2}(s,t)) + \\ &+ \cdots + D(F_1(s,t), F_0(s,t)) \leq (B^{m-1} + B^{m-2} + \cdots + 1) D^*(F_1, F_0) \leq \\ &\leq \frac{1}{1-B} D^*(F_1, F_0) \leq \frac{1}{1-B} (B\|g\|_{\mathscr{F}} + \|f\|_{\mathscr{F}} + N_K \Delta \|H\|_{\mathscr{F}}). \end{aligned}$$

Consequently for all $(s,t) \in A$ and $m \in N$ we have

$$\begin{aligned} &D(F_m(s,t),\tilde{0}) \leq D(F_m(s,t), g(s,t)) + D(g(s,t),\tilde{0}) \leq \\ &\leq \tfrac{1}{1-B}(B\|g\|_{\mathscr{F}} + \|f\|_{\mathscr{F}} + N_K \Delta \|H\|_{\mathscr{F}}) + \|g\|_{\mathscr{F}}. \end{aligned} \tag{9}$$

That is the uniformly bounded of sequence $\{F_m\}_{m \in N}$ in X.

For $m \in N$, let $U_m : A \to \mathbf{R}_{\mathscr{F}}$, $U_m(s,t) = H(s,t,F_m(s,t))$. Then

$$D(U_m(s,t),\tilde{0}) \leq D(H(s,t,F_m(s,t)),H(s,t,g(s,t))) + D(H(s,t,g(s,t)),\tilde{0}) \leq$$
$$\leq \alpha_2 D(F_m(s,t),g(s,t)) + \alpha_2 \|g\|_{\mathscr{F}} + \|H\|_{\mathscr{F}}.$$

Hence the sequence $\{U_m\}_{m\in N}$ is uniformly bounded in X. In addition, for any $(s,t) \in A$ and conditions (6) and (9) we have

$$D(F(s,t),\tilde{0}) \leq D(F(s,t),F_m(s,t)) + D(F_m(s,t),\tilde{0}) \leq$$
$$\leq \frac{1}{1-B}(B\|g\|_{\mathscr{F}} + \|f\|_{\mathscr{F}} + N_K \Delta \|H\|_{\mathscr{F}}) + \|g\|_{\mathscr{F}}.$$

We conclude that the solution of (1) is bounded. □

4 The Error Estimation

We present a numerical method to solve the Eq. (1) and define uniform partitions $a = a_0 < a_1 < \cdots < a_n = b$ and $c = c_0 < c_1 < \cdots < c_n = d$, with intermediates points $\xi_i \in [a_{i-1},a_i]$ and $\eta_j \in [c_{j-1},c_j]$, $i = \overline{1,n}$;$j = \overline{1,n}$, $h = \frac{b-a}{n}$, $h' = \frac{d-c}{n}$. Then the following iterative procedure given the approximate solution of Eq. (1) in point $(s,t) \in A$, $m = 1,2,...,$

$$\begin{aligned} &\tilde{F}_0(s,t) = g(s,t), \quad \tilde{F}_m(s,t) = g(s,t) \oplus f(s,t,\tilde{F}_{m-1}(s,t)) \oplus \\ &\oplus hh' \sum_{j=1}^{n} \sum_{i=1}^{n} K(s,t,\xi_i,\eta_j) \odot H(\xi_i,\eta_j,\tilde{F}_{m-1}(\xi_i,\eta_j)). \end{aligned} \tag{10}$$

Lemma 3 *Under the conditions (i)-(iv) we have*

$$\omega_A(F_m,hh') \leq \frac{1}{1-\alpha_1}\omega_A(g,hh') + \frac{\gamma_1}{1-\alpha_1}(h+h') + \frac{\Delta(\alpha_2\Gamma + \|H\|_{\mathscr{F}})}{1-\alpha_1}\omega_1,$$

where $\Gamma = \max_{0\leq i\leq m-1} \|F_i\|_{\mathscr{F}}$ *and for all* $\delta > 0$, $\omega_1 = \omega_A(K,\delta) =$ $= \sup_{(s_i,t_i)\in A; i=1,2} \{|K(s_1,t_1,x,y) - K(s_2,t_2,x,y)| : \sqrt{(s_1-s_2)^2 + (t_1-t_2)^2} \leq \delta\}.$

Proof Under for $(x_1,y_1),(x_2,y_2) \in A$ with $\sqrt{(x_1-x_2)^2 + (y_1-y_2)^2} \leq hh'$, by using Lemmas 1, 2 and (9) it obtains

$$D(F_m(x_1,y_1),F_m(x_2,y_2)) \leq D(g(x_1,y_1),g(x_2,y_2)) +$$
$$+ D(f(x_1,y_1,F_{m-1}(x_1,y_1)),f(x_2,y_2,F_{m-1}(x_2,y_2)) +$$

$$+D((FR)\int_c^d (FR)\int_a^b K(x_1,y_1,x,y)\odot H(x,y,F_{m-1}(x,y))dxdy,$$

$$(FR)\int_c^d (FR)\int_a^b K(x_2,y_2,x,y)\odot H(x,y,F_{m-1}(x,y))dxdy))\leq$$

$$\leq \omega_A(g,hh')+\gamma_1(h+h')+\alpha_1 D(F_{m-1}(x_1,y_1),F_{m-1}(x_2,y_2))+$$

$$+\int_c^d\int_a^b |K(x_1,y_1,x,y)-K(x_2,y_2,x,y)|D(H(x,y,F_{m-1}(x,y)),\tilde{0})dxdy\leq$$

$$\leq \omega_A(g,hh')+\gamma_1(h+h')+\alpha_1 D(F_{m-1}(x_1,y_1),F_{m-1}(x_2,y_2))+$$

$$+\omega_1\int_c^d\int_a^b (\alpha_2 D(F_{m-1}(x,y),\tilde{0})+D(H(x,y,\tilde{0}),\tilde{0}))dxdy\leq$$

$$\leq \omega_A(g,hh')+\gamma_1(h+h')+\alpha_1 D(F_{m-1}(x_1,y_1),F_{m-1}(x_2,y_2))+$$
$$+\omega_1\Delta(\alpha_2\|F_{m-1}\|_{\mathscr{F}}+\|H\|_{\mathscr{F}})\leq P+\alpha_1 D(F_{m-1}(x_1,y_1),F_{m-1}(x_2,y_2))+\omega_1\Delta\alpha_2\|F_{m-1}\|_{\mathscr{F}},$$

where $P=\omega_A(g,hh')+\gamma_1(h+h')+\omega_1\Delta\|H\|_{\mathscr{F}}$.
So, we have

$$D(F_m(x_1,y_1),F_m(x_2,y_2))\leq P+\alpha_1 D(F_{m-1}(x_1,y_1),F_{m-1}(x_2,y_2))+\omega_1\Delta\alpha_2\|F_{m-1}\|_{\mathscr{F}},$$
$$D(F_{m-1}(x_1,y_1),F_{m-1}(x_2,y_2))\leq P+\alpha_1 D(F_{m-2}(x_1,y_1),F_{m-2}(x_2,y_2))+\omega_1\Delta\alpha_2\|F_{m-2}\|_{\mathscr{F}},$$
$$\ldots$$
$$D(F_1(x_1,y_1),F_1(x_2,y_2))\leq P+\alpha_1 D(F_0(x_1,y_1),F_0(x_2,y_2))+\omega_1\Delta\alpha_2\|F_0\|_{\mathscr{F}}.$$

Multiplying these inequalities by $1,\alpha_1,\ldots,\alpha_1^{m-1}$, respectively, and summing them we have

$$D(F_m(x_1,y_1),F_m(x_2,y_2))\leq(1+\alpha_1+\cdots+\alpha_1^{m-1})P+\alpha_1^m\omega_A(g,hh')+$$
$$+\omega_1\Delta\alpha_2(\|F_{m-1}\|_{\mathscr{F}}+\alpha_1\|F_{m-2}\|_{\mathscr{F}}+\cdots+\alpha_1^{m-1}\|F_0\|_{\mathscr{F}})\leq$$
$$\leq\frac{1}{1-\alpha_1}\omega_A(g,hh')+\frac{1}{1-\alpha_1}(\gamma_1(h+h')+\omega_1\Delta\|H\|_{\mathscr{F}})+\frac{1}{1-\alpha_1}\omega_1\Delta\alpha_2\Gamma.$$

□

Theorem 4 *Under the conditions (i)-(iv) the iterative method (10) converges to the unique solution F of (1) and its error estimate is as follows*

$$D^*(F,\tilde{F}_m)\leq\frac{B^m}{1-B}(B\|F_0\|_{\mathscr{F}}+\|f\|_{\mathscr{F}}+N_K\Delta\|H\|_{\mathscr{F}})+$$

$$+\frac{5\Delta N_K}{4(1-B)}(\gamma_2+\frac{\alpha_2\gamma_1}{1-\alpha_1})(h+h')+\frac{5\Delta N_K\alpha_2}{4(1-\alpha_1)(1-B)}\omega_A(g,hh')+$$

$$+\frac{5\Delta^2 N_K \alpha_2(\alpha_2\Gamma + \|H\|_{\mathscr{F}})}{4(1-\alpha_1)(1-B)}\omega_1 + \frac{3\Delta(\alpha_2\mu + \|H\|_{\mathscr{F}})}{2(1-B)}\omega_2,$$

where Γ, ω_1 *are of Lemma 2,* $\mu = \max_{0\le i\le m-1} \|\tilde{F}_i\|_{\mathscr{F}}$ *and for all* $\delta > 0$, $\omega_2 = \omega_A(K,\delta) =$

$$= \sup_{(x_i,y_i)\in A, i=1,2} \{|K(s,t,x_1,y_1) - K(s,t,x_2,y_2)| : \sqrt{(x_1-x_2)^2 + (y_1-y_2)^2} \le \delta\}.$$

Proof Considering iterative procedure (10), for all $(s,t) \in A$ we have

$$D(F_m(s,t), \tilde{F}_m(s,t)) = D(g(s,t), g(s,t)) + D(f(s,t,F_{m-1}(s,t)), f(s,t,\tilde{F}_{m-1}(s,t)))+$$

$$+ D((FR)\int_c^d (FR)\int_a^b K(s,t,x,y)\odot H(x,y,F_{m-1}(x,y))dxdy,$$

$$hh'\sum_{j=1}^n\sum_{i=1}^n K(s,t,\xi_i,\eta_j)\odot H(\xi_i,\eta_j,\tilde{F}_{m-1}(\xi_i,\eta_j))) \le$$

$$\le \alpha_1 D^*(F_{m-1},\tilde{F}_{m-1})+$$

$$+\sum_{j=1}^n\sum_{i=1}^n D((FR)\int_{c_{j-1}}^{c_j}(FR)\int_{a_{i-1}}^{a_i} K(s,t,x,y)\odot H(x,y,F_{m-1}(x,y))dxdy,$$

$$hh'K(s,t,x,y)\odot H(\xi_i,\eta_j,F_{m-1}(\xi_i,\eta_j)))+$$

$$+\sum_{j=1}^n\sum_{i=1}^n D(hh'K(s,t,x,y)\odot H(\xi_i,\eta_j,F_{m-1}(\xi_i,\eta_j)),$$

$$hh'K(s,t,x,y)\odot H(\xi_i,\eta_j,\tilde{F}_{m-1}(\xi_i,\eta_j)))+$$

$$+\sum_{j=1}^n\sum_{i=1}^n D(hh'K(s,t,x,y)\odot H(\xi_i,\eta_j,\tilde{F}_{m-1}(\xi_i,\eta_j)),$$

$$hh'K(s,t,\xi_i,\eta_j)\odot H(\xi_i,\eta_j,\tilde{F}_{m-1}(\xi_i,\eta_j))) \le$$

$$\le \alpha_1 D^*(F_{m-1},\tilde{F}_{m-1}) + \frac{5}{4}\Delta N_K\omega_A(H(x,y,F_{m-1}(x,y)),hh') + \Delta N_K\alpha_2 D^*(F_{m-1},\tilde{F}_{m-1})+$$

$$+\frac{3}{2}hh'\omega_2\sum_{j=1}^n\sum_{i=1}^n(D(H(\xi_i,\eta_j,\tilde{F}_{m-1}(\xi_i,\eta_j)),H(\xi_i,\eta_j,\tilde{0})) + D(H(\xi_i,\eta_j,\tilde{0}),\tilde{0})) \le$$

$$\le BD^*(F_{m-1},\tilde{F}_{m-1}) + \frac{5\Delta N_K}{4}\omega_A(H(x,y,F_{m-1}(x,y)),hh') + \frac{3\Delta\omega_2(\alpha_2\|\tilde{F}_{m-1}\|_{\mathscr{F}} + \|H\|_{\mathscr{F}})}{2}.$$

For $\omega_A(H(x,y,F_{m-1}(x,y)),hh')$ we obtain

$$D(H(x_1,y_1,F_{m-1}(x_1,y_1)),H(x_2,y_2,F_{m-1}(x_2,y_2))) \le$$

$$\le \gamma_2(h+h') + \alpha_2 D(F_{m-1}(x_1,y_1),F_{m-1}(x_2,y_2)) \le \gamma_2(h+h') + \alpha_2\omega_A(F_{m-1},hh').$$

From Lemma 3 we obtain

$$D^*(F_m, \tilde{F}_m) \leq BD^*(F_{m-1}, \tilde{F}_{m-1}) + \frac{5\Delta N_K \gamma_2}{4}(h+h') + \frac{5\Delta N_K \alpha_2}{4(1-\alpha_1)}\omega_A(g, hh')+$$
$$+\frac{5\Delta N_K \alpha_2 \gamma_1}{4(1-\alpha_1)}(h+h') + \frac{5\Delta^2 N_K \alpha_2(\alpha_2\Gamma + \|H\|_{\mathscr{F}})}{4(1-\alpha_1)}\omega_1 + \frac{3}{2}\Delta\omega_2\alpha_2\|\tilde{F}_{m-1}\|_{\mathscr{F}} + \frac{3\Delta\|H\|_{\mathscr{F}}}{2}\omega_2.$$

We denote

$$Q = \frac{5\Delta N_K}{4}(\gamma_2 + \frac{\alpha_2\gamma_1}{1-\alpha_1})(h+h') + \frac{5\Delta N_K \alpha_2}{4(1-\alpha_1)}\omega_A(g, hh')+$$
$$+\frac{5\Delta^2 N_K \alpha_2(\alpha_2\Gamma + \|H\|_{\mathscr{F}})}{4(1-\alpha_1)}\omega_1 + \frac{3\Delta\|H\|_{\mathscr{F}}}{2}\omega_2.$$

Hence we conclude

$$D^*(F_m, \tilde{F}_m) \leq BD^*(F_{m-1}, \tilde{F}_{m-1}) + \frac{3}{2}\Delta\omega_2\alpha_2\|\tilde{F}_{m-1}\|_{\mathscr{F}} + Q$$
$$D^*(F_{m-1}, \tilde{F}_{m-1}) \leq BD^*(F_{m-2}, \tilde{F}_{m-2}) + \frac{3}{2}\Delta\omega_2\alpha_2\|\tilde{F}_{m-2}\|_{\mathscr{F}} + Q$$
$$\dots$$
$$D^*(F_1, \tilde{F}_1) \leq BD^*(F_0, \tilde{F}_0) + \frac{3}{2}\Delta\omega_2\alpha_2\|\tilde{F}_0\|_{\mathscr{F}} + Q.$$

Multiplying these inequalities by 1, B, ..., B^{m-1} respectively and summing them obtain

$$D^*(F_m, \tilde{F}_m) \leq \frac{3}{2}\Delta\omega_2\alpha_2(\|\tilde{F}_{m-1}\|_{\mathscr{F}} + B\|\tilde{F}_{m-2}\|_{\mathscr{F}} + \cdots \} + B^{m-1}\|\tilde{F}_0\|_{\mathscr{F}})+$$
$$+Q(1 + B + \cdots + B^{m-1}) \leq \frac{3}{2(1-B)}\Delta\omega_2\alpha_2\mu + Q\frac{1}{1-B}.$$

Hence we obtain

$$D^*(F_m, \tilde{F}_m) \leq \frac{5\Delta N_K}{4(1-B)}(\gamma_2 + \frac{\alpha_2\gamma_1}{1-\alpha_1})(h+h') + \frac{5\Delta N_K \alpha_2}{4(1-\alpha_1)(1-B)}\omega_A(g, hh')+$$
$$+\frac{5\Delta^2 N_K \alpha_2(\alpha_2\Gamma + \|H\|_{\mathscr{F}})}{4(1-\alpha_1)(1-B)}\omega_1 + \frac{3\Delta(\alpha_2\mu + \|H\|_{\mathscr{F}})}{2(1-B)}\omega_2. \quad (11)$$

From condition (6) of Theorem 3 we obtained

$$D^*(F,\tilde{F}_m) \le D^*(F,F_m) + D^*(F_m,\tilde{F}_m) \le \frac{B^m}{1-B}(B\|g\|_{\mathscr{F}} + \alpha_1\|f\|_{\mathscr{F}} + N_K\Delta\|H\|_{\mathscr{F}})+$$

$$+\frac{5\Delta N_K}{4(1-B)}(\gamma_2 + \frac{\alpha_2\gamma_1}{1-\alpha_1})(h+h') + \frac{5\Delta N_K\alpha_2}{4(1-\alpha_1)(1-B)}\omega_A(g,hh')+$$

$$+\frac{5\Delta^2 N_K\alpha_2(\alpha_2\Gamma + \|H\|_{\mathscr{F}})}{4(1-\alpha_1)(1-B)}\omega_1 + \frac{3\Delta(\alpha_2\mu + \|H\|_{\mathscr{F}})}{2(1-B)}\omega_2.$$

□

Remark 1 Since $B < 1$, $\lim_{h,h'\to 0}\omega_A(g,hh') = 0$, $\lim_{h,h'\to 0}\omega_A(K,hh') = 0$ and $\lim_{h,h'\to 0}\omega_A(K,h+h') = 0$, is to prove that $\lim_{m\to\infty,h,h'\to 0} D^*(F,\tilde{F}_m) = 0$ that shows the convergence of the method.

5 Numerical Stability Analysis

We study the numerical stability of the iterative algorithm (4) with respect to small changes in the starting approximation. We consider $F_0 = g$ and another starting approximation $G_0 = g^* \in C(A,\mathbf{R}_{\mathscr{F}})$ such that exists $\varepsilon > 0$ for which $D(F_0(s,t), G_0(s,t)) < \varepsilon$, for all $(s,t)\in A$. The obtained sequence of successive approximations is:

$$\begin{aligned}
&G_0(s,t) = g^*(s,t),\\
&G_m(s,t) = g(s,t) \oplus f(s,t,G_{m-1}(s,t)) \oplus\\
&\oplus (FR)\int_c^d (FR)\int_a^b K(s,t,x,y)\odot H(x,y,G_{m-1}(x,y))dxdy, m = 1,2,...,
\end{aligned}$$

and using the iterative method, the term of produced sequence are:

$$\begin{aligned}
&\tilde{G}_0(s,t) = g^*(s,t),\\
&\tilde{G}_m(s,t) = g(s,t) \oplus f(s,t,\tilde{G}_{m-1}(s,t)) \oplus\\
&\oplus hh'\sum_{j=1}^{n}\sum_{i=1}^{n} K(s,t,\xi_i,\eta_j)\odot H(\xi_i,\eta_j,\tilde{G}_{m-1}(\xi_i,\eta_j)),\ m = 1,2,...
\end{aligned}$$

Definition 7 The method of successive approximations applied to the integral Eq. (1) is said to be numerically stable with respect to the choice of the first iteration if for all $(s,t) \in A$ there exist constants $k_1, k_2, k_3, k_4, k_5, k_6 > 0$ which are independent by $h = \frac{b-a}{n}$ and $h' = \frac{d-c}{n}$ such that

$$D(\tilde{F}_m(s,t), \tilde{G}_m(s,t)) < k_1\varepsilon + k_2(h+h') + k_3\omega_A(g, hh') + k_4\omega_A(g^*, hh') + k_5\omega_1 + k_6\omega_2.$$

Theorem 5 *Under the conditions (i)-(iv) the iterative method is numerically stable with respect to the choice of the first iteration.*

Proof First, we observe that
$D^*(\tilde{F}_m, \tilde{G}_m) \le D^*(\tilde{F}_m, F_m) + D^*(F_m, G_m) + D^*(G_m, \tilde{G}_m)$.
From inequality (11) of Theorem 3 we have

$$D^*(G_m, \tilde{G}_m) \le \frac{5\Delta N_K}{4(1-B)}(\gamma_2 + \frac{\alpha_2\gamma_1}{1-\alpha_1})(h+h') + \frac{5\Delta N_K\alpha_2}{4(1-\alpha_1)(1-B)}\omega_A(g^*, hh') +$$
$$+\frac{5\Delta^2 N_K\alpha_2(\alpha_2\Gamma^* + \|H\|_{\mathcal{F}})}{4(1-\alpha_1)(1-B)}\omega_1 + \frac{3\Delta(\alpha_2\mu^* + \|H\|_{\mathcal{F}})}{2(1-B)}\omega_2,$$

where $\Gamma^* = \max_{0\le i\le m-1} \|G_i\|_{\mathcal{F}}$, $\mu^* = \max_{0\le i\le m-1} \|\tilde{G}_i\|_{\mathcal{F}}$.

By hypothesis, $D(F_0(s,t), G_0(s,t)) < \varepsilon$, for all $(s,t) \in A$ and thus

$$D(F_m(s,t), G_m(s,t)) \le D(g(s,t), g(s,t)) + D(f(s,t,F_{m-1}(s,t)), f(s,t,G_{m-1}(s,t))) +$$
$$+ D((FR)\int_c^d (FR)\int_a^b K(s,t,x,y) \odot H(x,y,F_{m-1}(x,y))dxdy,$$
$$(FR)\int_c^d (FR)\int_a^b K(s,t,x,y) \odot H(x,y,G_{m-1}(x,y))dxdy) \le \alpha_1 D^*(F_{m-1}, G_{m-1}) +$$
$$+ (FR)\int_c^d (FR)\int_a^b |K(s,t,x,y)| D(H(x,y,F_{m-1}(x,y)), H(x,y,G_{m-1}(x,y)))dxdy) \le$$
$$\le \alpha_1 D^*(F_{m-1}, G_{m-1}) + N_K\Delta\alpha_2 D^*(F_{m-1}, G_{m-1}) = BD^*(F_{m-1}, G_{m-1}).$$

Then $D^*(F_m, G_m) \le B^m D^*(F_0, G_0) \le B^m\varepsilon$ for all $(s,t) \in A$, $m \ge 1$ and
$D^*(\tilde{F}_m, \tilde{G}_m) \le k_1\varepsilon + k_2(h+h') + k_3\omega_A(g, hh') + k_4\omega_A(g^*, hh') + k_5\omega_1 + k_6\omega_2$,

were $k_1 = B^m$, $k_2 = \frac{5\Delta N_K}{2(1-B)}(\gamma_2 + \frac{\alpha_2\gamma_1}{1-\alpha_1})$, $k_3 = k_4 = \frac{5\Delta N_K\alpha_2}{4(1-\alpha_1)(1-B)}$,

$k_5 = \frac{5\Delta^2 N_K\alpha_2(\alpha_2(\Gamma+\Gamma^*) + 2\|H\|_{\mathcal{F}})}{4(1-\alpha_1)(1-B)}$, $k_6 = \frac{3\Delta(\alpha_2(\mu+\mu^*) + 2\|H\|_{\mathcal{F}})}{2(1-B)}$. □

6 Numerical Experiments

In this section, we intent to illustrate the obtained theoretical results on some numerical example testing the convergence of the method and the numerical stability with respect to the choice of the first iteration. The algorithm was implemented using C♯. The program can be found on the following web address: http://math.asm32.info/r/fuzzy.

Example Let $A = [0, 1] \times [0, 1]$. For the integral equation

$$F(s,t) = g(s,t) \oplus f(s,t,F(s,t)) \oplus (FR)\int_c^d (FR)\int_a^b stxy \odot F^2(x,y)dxdy,\ (s,t) \in A$$

the exact solution is $\underline{F}(s,t,r) = (2+r)st$ and $\bar{F}(s,t,r) = (4-r)st$.

Here $\underline{g}(s,t,r) = \frac{3}{4}(2+r)st - \frac{1}{8}(2+r)^2 st$, $\bar{g}(s,t,r) = \frac{3}{4}(4-r)st - \frac{1}{8}(4-r)^2 st$,

$\underline{f}(s,t,\underline{F}(s,t,r),r) = \frac{\underline{F}(s,t,r)}{4} + \frac{1}{16}(2+r)^2 st$, $\bar{f}(s,t,\bar{F}(s,t,r),r) = \frac{\bar{F}(s,t,r)}{4} + \frac{1}{16}(4-r)^2 st$. Applying the iterative algorithm for various m, n, $r_i = ih_r$, $i = 0 \div 10$, $h_r = \frac{1}{10}$ we obtain the computational errors $\underline{E}_m(r_i) = \underline{E}_m(s_0,t_0,r_i) = |\underline{\tilde{F}}_m(s_0,t_0,r_i) - \underline{F}(s_0,t_0,r_i)|$ and $\bar{E}_m(r_i) = \bar{E}_m(s_0,t_0,r_i) = |\bar{\tilde{F}}_m(s_0,t_0,r_i) - \bar{F}(s_0,t_0,r_i)|$ in the point $(s_0,t_0) = (0.5, 0.5)$. We get that for any $0 \leq r \leq 1$, the norm of the errors tend to zero as $m, n \to \infty$ we present these results in Table 1. The numerical stability is tested by considering $\varepsilon = 0.1$, and for various m, n. The results are expressed by $\underline{D}_m(r_i) = \underline{D}_m(s_0,t_0,r_i) = |\underline{\tilde{F}}_m(s_0,t_0,r_i) - \underline{\tilde{G}}_m(s_0,t_0,r_i)|$ and $\bar{D}_m(r_i) = \bar{D}_m(s_0,t_0,r_i) = |\bar{\tilde{F}}_m(s_0,t_0,r_i) - \bar{\tilde{G}}_m(s_0,t_0,r_i)|$ in the point $(s_0,t_0) = (0.5, 0.5)$ in Table 2.

Table 1 Numerical errors in (0.5, 0.5)

	m = 10 n = 10		m = 20 n = 10		m = 10 n = 30		m = 20 n = 30	
r_i	$\underline{E}_m(r_i)$	$\bar{E}_m(r_i)$	$\underline{E}_m(r_i)$	$\bar{E}_m(r_i)$	$\underline{E}_m(r_i)$	$\bar{E}_m(r_i)$	$\underline{E}_m(r_i)$	$\bar{E}_m(r_i)$
0	1,43E-03	2,77E-02	1,25E-03	2,11E-02	3,29E-04	1,06E-02	1,39E-04	2,19E-03
0,1	1,66E-03	2,39E-02	1,42E-03	9,50E-03	4,15E-04	1,78E-02	1,57E-04	1,77E-03
0,2	1,92E-03	2,07E-02	1,59E-03	8,51E-03	5,23E-04	1,49E-02	1,77E-04	1,45E-03
0,3	2,22E-03	1,78E-02	1,79E-03	7,64E-03	6,59E-04	1,24E-02	1,99E-04	1,21E-03
0,4	2,57E-03	1,54E-02	2,00E-03	6,88E-03	8,28E-04	1,03E-02	2,24E-04	1,01E-03
0,5	2,97E-03	1,32E-02	2,23E-03	6,21E-03	1,04E-03	8,56E-03	2,50E-04	8,62E-04
0,6	3,44E-03	1,14E-02	2,48E-03	5,61E-03	1,30E-03	7,06E-03	2,80E-04	7,41E-04
0,7	3,99E-03	9,80E-03	2,76E-03	5,07E-03	1,63E-03	5,80E-03	3,13E-04	6,44E-04
0,8	4,62E-03	8,42E-03	3,06E-03	4,59E-03	2,03E-03	4,74E-03	3,50E-04	5,64E-04
0,9	5,37E-03	7,24E-03	3,39E-03	4,15E-03	2,53E-03	3,86E-03	3,92E-04	4,97E-04
1	6,23E-03	6,23E-03	3,75E-03	3,75E-03	3,13E-03	3,13E-03	4,41E-04	4,41E-04

Table 2 Numerical errors in (0.5, 0.5)

	m = 10 n = 10		m = 20 n = 10		m = 10 n = 30		m = 20 n = 30	
r_i	$\underline{D}_m(r_i)$	$\overline{D}_m(r_i)$	$\underline{D}_m(r_i)$	$\overline{D}_m(r_i)$	$\underline{D}_m(r_i)$	$\overline{D}_m(r_i)$	$\underline{D}_m(r_i)$	$\overline{D}_m(r_i)$
0	2,63E-05	4,77E-04	2,40E-08	2,26E-05	2,75E-05	5,12E-04	2,70E-08	2,70E-05
0,1	3,24E-05	4,38E-04	3,80E-08	1,77E-05	3,39E-05	4,69E-04	4,20E-08	2,11E-05
0,2	3,96E-05	3,99E-04	5,90E-08	1,38E-05	4,15E-05	4,27E-04	6,50E-08	1,64E-05
0,3	4,80E-05	3,62E-04	9,00E-08	1,06E-05	5,04E-05	3,87E-04	1,00E-07	1,25E-05
0,4	5,78E-05	3,26E-04	1,36E-07	8,07E-06	6,08E-05	3,48E-04	1,52E-07	9,50E-06
0,5	6,91E-05	2,92E-04	2,03E-07	6,08E-06	7,28E-05	3,11E-04	2,28E-07	7,13E-06
0,6	8,21E-05	2,60E-04	2,99E-07	4,54E-06	8,66E-05	2,77E-04	3,38E-07	5,30E-06
0,7	9,70E-05	2,30E-04	4,37E-07	3,35E-06	1,02E-04	2,45E-04	4,96E-07	3,90E-06
0,8	1,14E-04	2,02E-04	6,30E-07	2,45E-06	1,20E-04	2,15E-04	7,18E-07	2,84E-06
0,9	1,33E-04	1,77E-04	8,99E-07	1,77E-06	1,40E-04	1,88E-04	1,03E-06	2,04E-06
1	1,53E-04	1,53E-04	1,27E-06	1,27E-06	1,63E-04	1,63E-04	1,46E-06	1,46E-06

Acknowledgements Research was partially supported by Fund FP17-FMI-008, Fund Scientific Research, University of Plovdiv Paisii Hilendarski.

References

1. Anastassiou, G.A.: Fuzzy Mathematics: Approximation Theory. Springer, Berlin (2010)
2. Balachandran, K., Kanagarajan, K.: Existence of solutions of general nonlinear fuzzy Volterra-Fredholm integral equations. J. Appl. Math. Stoch. Anal. **3**, 333–343 (2005)
3. Bede, B., Gal, S.: Quadrature rules for integrals of fuzzy-number-valued functions. Fuzzy Sets Syst. **145**, 359–380 (2004)
4. Buckley, J., Eslami, E., Feuring, T.: Fuzzy integral equations. In: Fuzzy Mathematics in Economics and Engineering. Studies in Fuzziness and Soft Computing, vol. 91, pp. 229–241. Springer, Physica-Verlag, Heidelberg (2002)
5. Dubois, D., Prade, H.: Towards fuzzy differential calculus. Part 2: integration of fuzzy intervals. Fuzzy Sets Syst. **8**, 105–116 (1982)
6. Fard, O.S., Sanchooli, M.: Two successive schemes for numerical solution of linear fuzzy Fredholm integral equations of the second kind. Aust. J. Appl. Sci. **4**, 817–825 (2010)
7. Friedman, M., Ma, M., Kendal, A.: Solutions to fuzzy integral equations with arbitrary kernels. Int. J. Approx. Reason **20**, 249–262 (1999)
8. Goetschel, R., Voxman, W.: Elementary fuzzy calculus. Fuzzy Sets Syst. **18**, 31–43 (1986)
9. Kaleva, O.: Fuzzy differential equations. Fuzzy Sets Syst. **24**, 301–317 (1987)
10. Park, J.Y., Jeong, J.U.: On the existence and uniqueness of solutions of fuzzy VoltteraFredholm integral equations. Fuzzy Sets Syst. **115**, 425–431 (2000)
11. Sadatrasoul, S.M., Ezzati, R.: Quadrature rules and iterative method for numerical solution of two-dimensional fuzzy integral equations. Abstr. Appl. Anal. **2014** (2014). doi:10.1155/2014/413570
12. Sadatrasoul, S.M., Ezzati, R.: Numerical solution of two-dimensional nonlinear Hammerstein fuzzy integral equations based on optimal fuzzy quadrature formula. J. Comput. Appl. Math. **292**, 430–446 (2016)
13. Seikkala, S.: On the fuzzy initial value problem. Fuzzy Sets Syst. **24**, 319–330 (1987)
14. Wu, C., Gong, Z.: On Henstock integral of fuzzy-number-valued functions (1). Fuzzy Sets Syst. **120**, 523–532 (2001)

Noise Removal and Feature Extraction of 2D CT Radiographic Images

Stanislav Harizanov, Jaume de Dios Pont, Sebastian Ståhl and Dennis Wenzel

Abstract 2D CT radiographic images are widely used in industrial as well as medical applications to examine different types of objects whenever non-destructive measurements of quality are necessary. To extract meaningful structural information for the scanned object from a low-dose input without increasing the radiation level of the scanner, we propose and experimentally investigate a novel two-step process. Firstly, the image is denoised by a regularization method in order to remove unwanted disturbances which affect its quality. Secondly, the difference images between the outputs of different regularization methods are used for feature localization and extraction. The theory as well as the numerical results of the application of several methods on real-life industrial CT data are presented and compared herein.

Keywords Image denoising · Feature extraction · Edge detection · Radiographic images · CT data reconstruction

---

S. Harizanov
Institute of Information and Communication Technologies,
Bulgarian Academy of Sciences, Sofia, Bulgaria
e-mail: sharizanov@parallel.bas.bg; sharizanov@math.bas.bg

S. Harizanov
Institute of Mathematics and Informatics, Bulgarian Academy of Sciences,
Sofia, Bulgaria

J. de Dios Pont
University of California Los Angeles, Los Angeles, CA, USA

S. Ståhl
Chalmers University of Technology, Göteborg, Sweden

D. Wenzel (✉)
Institute of Numerical Analysis, Dresden University of Technology,
Dresden, Germany
e-mail: dennis.wenzel@tu-dresden.de

K. Georgiev et al. (eds.), *Advanced Computing in Industrial Mathematics*, Studies in Computational Intelligence 728,
https://doi.org/10.1007/978-3-319-65530-7_6

1 Introduction

Image noise is a common problem in many applications. In this context noise is defined to be a disturbance in the observed signal, leading to an inaccurate measurement of the observed quantity and thus to a loss of information. There are various denoising techniques known and their effectiveness depends on how well the underlying processes of noise generation are investigated. An essential part in finding an appropriate denoising algorithm is the ability to accurately model and characterize the statistical parameters of the image noise distribution. This problem is addressed and solved for a single-frame series of 2D CT data. We provide numerical evidence that scaled Poisson-Gaussian noise model seems to be the right one for this setting.

The paper investigates different methods for 2D radiographic image denoising, based on regularization and convex optimization techniques. Those approaches typically give rise to over-smoothened results, since one chooses the most regular solution in a class of admissible images, containing the true noise-free reconstruction. We turn this drawback into an advantage and study the possibility to extract structural information for the scanned object from the segmented difference image of two differently denoised outputs. Since the noise part of the image is assumed to be the least regular one, as long as the noise-free image remains admissible, all the various outputs are expected to be noise-free and only their edge sharpness to be affected. Hence, the set of pixels that substantially change their gray-scale intensity level between two such outputs most likely belongs to the image edges and can be visualized in high-contrast via direct segmentation of their difference image.

One way of dealing with Poisson-dominated noise is to apply a variance-stabilizing transformation (VST). In this work the Anscombe transform is considered. It is developed in [1], transforms Poisson noise into a Gaussian one with zero mean and unit variance, and has been used to denoise photographs and digital images in [2, 3]. An alternative denoising approach is to directly solve the convex optimization problem of regularization while using the Maximum A Posteriori (MAP) estimate of the underlying noise as data fidelity measurement. The I-divergence functional is the neg-log likelihood estimator of the Poisson distribution and it can be effectively incorporated as constraint in the optimization [4].

The paper is organized as follows. Section 2 contains detailed statistical analysis on the parameters of the noise distribution for CT data. In Sect. 3 various denoising methods are introduced and summarized. In Sect. 4 the proposed procedure of using oversmoothing for feature extraction is explained. In Sect. 5 we perform numerical experiments on two real-world industrial examples. Conclusions are drawn in Sect. 6.

2 Statistical Analysis on Noise Characteristics

2.1 Estimation of Noise Parameters

In order to design efficient denoising algorithms we first need to understand and properly model the underlying process of noise generation. CT-scanners work by counting photons, so the usual assumption is that the dominant component of the generated noise is Poisson distributed. There is also an additive Gaussian noise, which is related to the data acquisition device and is supposed to describe the thermodynamic fluctuations. We assume that it is of zero mean and spatially uncorrelated, thus in this paper we investigate the mixed Poisson-Gaussian model

$$F = \alpha Poiss(\alpha^{-1}\bar{F}) + \mathcal{N}(0, \sigma^2). \tag{1}$$

The observed (noisy) image $F \in \mathbb{R}^{m \times n}$ is a realization of the right-hand-side distribution, $\bar{F}$ is the true (noise-free) image, while α and σ are assumed to be global parameters, independent of the pixel's position. All notations are to be understood componentwise. We have included the normalization constant α to compensate for the fact that commercial CT-scanners often use some type of scaling, i.e. the F gray-scale intensity values are not equal to the exact number of counted photons, but are only proportional to them.

Our goal is to derive reliable approximations of α and σ, which will be further used in the denoising process. We apply statistical tools, based on the relationship between the mean value $\bar{F}$ and the variance σ^2 in the images. For this purpose, we study series $\{F^{(k)}\}_{k=1}^{L}$ of radiographic images of the same object, generated in a narrow time frame under the same scanning conditions. Furthermore, we assume that each $F^{(k)}$ is an independent realization of (1), so we compute pixel-wise sample mean and variance via

$$\hat{\mu} = \frac{1}{L}\sum_{k=1}^{L} F^{(k)} \quad \text{and} \quad \hat{\sigma}^2 = \frac{1}{L}\sum_{k=1}^{L}(F^{(k)} - \hat{\mu})^2,$$

where we have that:

$$\hat{\mu} \sim \mathcal{N}\left(F, \frac{\sigma^2}{L}\right) \quad \text{and} \quad \hat{\sigma}^2 \sim \sigma^2 \frac{\chi^2_{L-1}}{L}.$$

We can then use this information to perform an MLE estimation for the parameters $\hat{\mu}$ and $\hat{\sigma}$, under the constraint $\hat{\sigma}^2 = \sigma^2 + \alpha\hat{\mu}$, implied by (1). In order to do so, it is suitable to perform the approximation $Poiss(k) \approx \mathcal{N}(k, k)$, which is reliable for large enough k (e.g., $k \geq 20$).

However, the variance and mean estimators we formulated so far are the ones observed at each point. They do not exactly satisfy the above constraint.

Therefore, we define two new estimators, $\tilde{\mu}$ and $\tilde{\sigma}^2$, which we obtain from a maximum likelihood estimation over all the points, imposing equation (1) as a condition between $\tilde{\mu}$ and $\tilde{\sigma}^2$. We write the likelihood function as:

$$l(\tilde{\mu}, \tilde{\sigma}^2) = -\frac{N}{2}\left(\sum_{i,j} \frac{\hat{\sigma}^2_{i,j} + (\hat{\mu}_{i,j} - \tilde{\mu}_{i,j})^2}{\tilde{\sigma}^2_{i,j}} - \log \tilde{\sigma}^2_{i,j}\right).$$

While there is no closed form for the global minimizer of l, it can be shown that $l(\tilde{\mu}, \tilde{\sigma}^2)$ is convex, thus, for example, gradient-descent-type methods will converge to it, under the relation $\tilde{\sigma}^2 = \sigma^2 + \alpha\tilde{\mu}$. A good seed for the gradient descent method can be obtained by initializing $(\tilde{\mu}, \tilde{\sigma}) = (\hat{\mu}, \hat{\sigma})$, and then solving the equation analytically. This leads to the best least squares fit

$$\begin{pmatrix} \langle \mathbf{1}_N, \mathbf{1}_N \rangle & \langle \tilde{\mu}, \mathbf{1}_N \rangle \\ \langle \mathbf{1}_N, \tilde{\mu} \rangle & \langle \tilde{\mu}, \tilde{\mu} \rangle \end{pmatrix} \begin{pmatrix} \sigma^2 \\ \alpha \end{pmatrix} = \begin{pmatrix} \langle \tilde{\sigma}^2, \mathbf{1}_N \rangle \\ \langle \tilde{\sigma}^2, \tilde{\mu} \rangle \end{pmatrix}. \tag{2}$$

Once α and σ are estimated, $\tilde{\mu}$ can be updated by equating the respective partial derivatives of l to 0 and solving the corresponding second degree equation, or by applying Newton's method. This gives rise to an efficient iterative method for the computation of the noise parameters, by sequentially updating the values of $\tilde{\mu}$ and (α, σ). The convexity of the log-likelihood guarantees the convergence of the method. Numerical experiments show convergence in 2 to 3 iterations of the method for series $\{F^{(k)}\}_{k=1}^{L}$ of size $L = 8, 16$.

The estimates for the first experimental dataset used in the paper are shown in Fig. 1. The orange line (moving average of the variance as a function of the mean value) shows a clear dependence between the variance and the mean value,

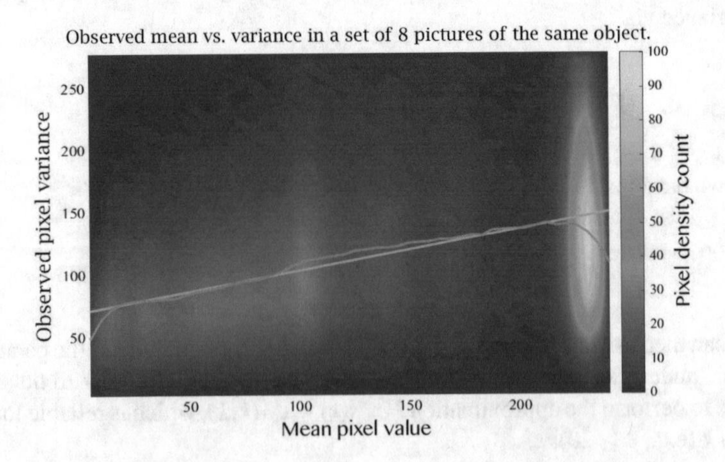

Fig. 1 Observed mean and variance for the 3D printed object. *Green* (*straight line*): estimate from the Poisson-Gaussian noise mixture. *Orange* (line with non-constant derivative): moving average for the noise

as hypothesized. At the two ends of the intensity interval, we observe larger deviations between the two lines, which is due to the restrictions on the gray-scale values to remain within the 8-bit bitmap format, thus the corresponding noise distributions there are "trimmed" and do not follow exactly (1). To avoid that, in practice we perform the statistical analysis only on the set of pixels whose gray-scale values lie inside of the intensity interval for all frames $F^{(k)}$, $k = 1, \dots, L$. In addition, we compute $\hat{\mu}$ from another, much larger series $\{\tilde{F}^{(k)}\}_{k=1}^{L}$, $L = 2048$. Therefore, we assume that $\hat{\mu}$ is a trustful approximation of the noise-free image $\bar{F}$, denote it by $avgF$ and solve the system (2) just once with $(\tilde{\mu}, \tilde{\sigma}) = (avgF, \hat{\sigma})$.

2.2 *Variance Stabilizing Transformation*

Once the noise parameters have been reliably estimated, we want to incorporate this information in our denoising process. Applying a proper Variance Stabilizing Transformation (VST) to the noisy data removes the mean-variance relation in the Poisson component and transforms the mixed Poisson-Gaussian noise into a purely Gaussian one. The Generalized Anscombe Transform is an example of a VST, for which the transformed noise is white (zero mean and variance one). It is given by (see Eq. 2.8 in [5])

$$T_G(F) = \frac{2}{\alpha}\sqrt{\alpha F + \frac{3\alpha^2}{8} + \sigma^2} \tag{3}$$

and

$$T_G : \alpha Poiss(\alpha^{-1}\bar{F}) + \mathcal{N}(0, \sigma^2) \mapsto \mathcal{N}(\bar{F}, 1).$$

Note that $\alpha = 1, \sigma = 0$ implies purely Poisson noise and is referred to as the regular Anscombe transform T.

The classical denoising approach for images, corrupted by mixed Poisson-Gaussian noise consists of: applying T_G and then removing the white noise from the transformed image. Hence, we also need the inverse transform which transforms the (denoised) image back to the original intensity domain. In this paper the exact unbiased inverse of the Generalized Anscombe transform (proposed in [6, (8)], and used by e.g. [2]) is used, which is defined as

$$T_G^{-1} = T_G^{-1}(F, \bar{F}) = \int_{-\infty}^{\infty} 2\sqrt{\alpha F + \frac{3\alpha^2}{8} + \sigma^2} \sum_{k=0}^{\infty} \left(\frac{\bar{F}^k e^{-\bar{F}}}{k!\sqrt{2\pi\sigma^2}} e^{-\frac{(F-k\alpha)^2}{2\sigma^2}} \right) dF.$$

Here, F and σ are again the observed pixel intensity and the variance of the Gaussian part respectively while $\bar{F}$ is the true noise-free image we want to recover. Note that $\bar{F}$ is not known a priori, therefore in our denoising procedures we replace it by the 2048-frame-averaged approximation $avgF$ of $\bar{F}$, discussed in the previous subsection.

3 Summary of the Used Denoising Techniques

This section is devoted to the introduction of the different denoising models we compared. In this paper we follow the regularization approach, meaning that we derive the denoised result as a solution of a certain convex optimization problem, involving a regularization term R and a data fidelity term DF. We restrict ourselves to discrete gradient regularizations, where $\nabla : \mathbb{R}^N \mapsto \mathbb{R}^{2N}$ consists of the forward finite differences between spatially neighboring pixels in horizontal and vertical direction. For the regularization term we consider either the ℓ^2 norm $\|\nabla \cdot\|_2$ of the gradient or the mixed $\ell^{2,1}$ norm of the gradient $\| \, |\nabla \cdot | \, \|_1$, also known as the Total Variation (TV) semi-norm. For the data fidelity term, we use the I-divergence operator, when working directly with the initial Poisson-dominated noise, and the ℓ^2 norm, when applying a VST to the input and working with pure Gaussian noise.

There are two main types of optimization problems, namely constrained and penalized. The constrained problems are of the form

$$\hat{F} = \arg\min R(\nabla F) \quad \text{subject to} \quad DF(F, \bar{F}) \leq \tau, \qquad \tau > 0, \tag{4}$$

and search for the most-regular image within a (convex!) constrained set. The penalized problems are of the form

$$\hat{F} = \arg\min R(\nabla F) + \lambda DF(F, \bar{F}), \qquad \lambda > 0, \tag{5}$$

and here the data fidelity term is directly incorporated in the cost function via a penalizer λ. For the problems we consider, the above two classes are equivalent, meaning that there is a one-to-one correspondence between τ and λ, such that the solutions of (4) and (5) coincide.

- **Continuous L^2 model**

The particular continuous case where both the data fidelity and the regularization term are taken to be their respective L^2 norms is analytically solvable. For this we assume that our image is represented by a function $f \in L^2(I)$ where $I \subset \mathbb{R}^2$ is a rectangle of size $m \times n$. We are looking for a smoothened image that is smooth enough to be in the Sobolev space $H^2(I)$:

$$\hat{F}_{L2} = \underset{f \in H^2(I)}{\arg\min}\{\Lambda_\lambda f\}, \qquad \Lambda_\lambda f = \|f - F\|_2^2 + \lambda \|\nabla f\|_2^2 \tag{6}$$

Minimization over all directional derivatives yields

$$0 = \frac{d}{dt}\Lambda_\lambda(\hat{F}_{L2} + th)\Big|_{t=0} = \langle \hat{F}_{L2} - F, h\rangle + \lambda \langle \nabla \hat{F}_{L2}, \nabla h\rangle, \qquad \forall h \in L^2(I).$$

It is reasonable to consider Dirichlet or Neumann boundary conditions, so we can invoke $\langle \nabla \hat{F}_{L2}, \nabla h \rangle = -\langle \Delta \hat{F}_{L2}, h \rangle$. As the above identity should hold for every h we conclude

$$\hat{F}_{L2} - \lambda \Delta \hat{F}_{L2} = F \tag{7}$$

This equation can be solved in the Fourier domain, where the above operator can be written as $-(k_x^2 + k_y^2)$:

$$(\hat{F}_{L2})\hat{} = \hat{F}\frac{1}{1 + \lambda(k_x^2 + k_y^2)}.$$

Now, by the convolution theorem, we can recover $\hat{F}_{L2}$ as

$$\hat{F}_{L2} = F * K, \qquad K = \sqrt{\frac{1}{8\pi\lambda^3}} K_0\left(\frac{r}{\sqrt{4\pi\lambda}}\right),$$

where K is the inverse Fourier transform of $\frac{1}{1+\lambda(k_x^2+k_y^2)}$, $r = \sqrt{(x^2 + y^2)}$ and K_0 is the first modified Bessel function of the second kind.

The derived solution, however, is numerically unstable since the derived kernel is singular at the origin. Therefore, it is not discretizable in a trivial way. Instead of this discretization two alternatives might be considered: either solve the discretized problem in the Fourier space, where there are no singularities to be discretized, or solve the differential equation (7) in its discretized version.

- **VST + BM3D filtering**

Let us consider the denoising model

$$\hat{F}_{BM3D,T_G} = T_G^{-1}(\Phi(T_G(F))) \tag{8}$$

where T_G is the Generalized Anscombe Transform and Φ represents some AWGN algorithm. We use Sect. 2.1 to approximate the noise parameters α and σ for the given numerical examples. For the AWGN denoising there exist many different algorithms, see e.g. [7–9]. The AWGN filter Φ we use is the so called BM3D denoising algorithm found in [7], which performs *collaborative filtering* on 3D arrays of similar 2D fragments of $T_G(F)$. The denoising technique (8) is studied in [2] and in our tests we use the BM3D implementation provided there.[1]

For grouping the 2D fragments, the pointwise ℓ^2-distance between blocks of identical size is used as a measure of dissimilarity. Thus, one can think of this approach as a regularization method, based on a *non-local* gradient $\|\nabla_{NL} T_G F\|_2$, related to the fragment grouping. We have run several numerical experiments on the model (6) in the transformed domain $T_G f$ and observed that there is no significant difference

[1] Available at http://www.cs.tut.fi/~foi/invansc/.

(both quantitatively and qualitatively) between the output of (6) and the minimizer of $\arg\min\{T_G^{-1}(\Lambda_\lambda(T_G(f)))\}$, meaning that $\|\nabla T_G F\|_2$ is practically proportional to $\|\nabla F\|_2$. With this in mind, we consider the model (8) as a non-local, constrained analogue of (6), where the constrained τ_{match}^{ht} is the maximal distance for which two blocks are considered similar (see [7] for details).

- **I-divergence constrained TV-minimization**

Consider the constrained problem:

$$\hat{F}_{I-div} = \arg\min_{f\in\mathbb{R}^{m\times n}} \| \, |\nabla f| \, \|_1 \quad \text{subject to} \quad D(F,f) \le \tau. \tag{9}$$

We choose the regularization term to be the TV-norm and the data fidelity term to be the *Kullback-Leibler divergence*:

$$D(F,f) = \begin{cases} \sum_{i,j} F_{i,j} \log\left(\frac{F_{i,j}}{f_{i,j}}\right) - F_{i,j} + f_{i,j} & \text{if } f_{i,j} > 0 \\ \infty & \text{otherwise} \end{cases}$$

It is well known that the Kullback-Leibler divergence (or I-divergence) is the negative logarithmic likelihood estimator of the Poisson distribution and thus it is a priori clear that this approach should work best with Poisson dominated noise. Denote by $N = mn$ the size of the image. In the case of pure Poisson noise, statistical arguments suggest that $\tau = N/2$ is the optimal choice for the constrained parameter, as $D(F,\bar{F}) \approx N/2$ and we choose the smallest constrained set, containing $\bar{F}$ with high probability. For mixed Poisson-Gaussian noise, there is no a priori estimation of τ, so in our numerical experiments, we take $\tau = D(F, avgF)$ as a straightforward generalization of the above argument. In order to assure positivity of the input, we denoise the image $\max(F, \mathbf{1})$.

For $F > 0$, $D(F,\cdot)$ is convex and the constraint $\{f : D(F,f) \le \tau\}$ is convex and non-empty for all $\tau \ge 0$. As the TV-norm is convex as well, problem (9) is a convex optimization problem, for which strong duality holds. Therefore, we can apply a suitable *Primal Dual Splitting* algorithm to it. In this paper, following [4], we deal with the Alternating Direction Method of Multipliers (ADMM). The algorithm uses three different primal-dual variable pairs to ensure the three key properties *regularization, data fidelity* and *non-negativity*.

- **Anscombe-constrained TV-minimization**

Denote by ν the maximal allowed gray-scale intensity (it is either 255 for 8-bit images, or 65535 for the 16-bit ones). Consider the optimization problem

$$\hat{F}_{TV} = \arg\min_{f\in[0,\nu]^N} \| \, |\nabla f| \, \|_1 \quad \text{subject to} \quad \|T_G(F) - T_G(f)\|_2^2 \le \tau. \tag{10}$$

Due to (1), $T_G(F) - T_G(\bar{F})$ should be a realization of $\mathcal{N}(0, 1)$, thus $\tau = N$ is the optimal choice for the constrained parameter. When we apply the pure Anscombe transform T instead of its generalization T_G, we use $\tau = \|T(F) - T(avgF)\|_2^2$. The pure Anscombe-constrained denoising model has been studied in [3]. We apply a modified version of the algorithm there, adapted to the T_G framework.

4 Feature Extraction via Denoising

When using regularization techniques for image denoising, typically we oversmooth the result. Indeed we are looking for the most regular solution within a class C of admissible images and in order to assure high level of noise removal we need the true noise-free image $\bar{F}$ to belong to this class. The image $\bar{F}$ is not known a priori and, as a result, the class C has to be quite broad. However, apart from the noise component, $\bar{F}$ contains edges and singularities that capture most of the structural information of the scanned object. Therefore, in practice there is always a more regular member $\tilde{F}$ of C and we cannot hope to completely reconstruct $\bar{F}$ from the input image F. Instead, we want to turn this drawback into an advantage and we propose a simple procedure for localization of the image details, based on the oversmoothing phenomenon. Consider two different classes C_1 and C_2 of images, such as with high probability $\bar{F} \in C_1 \cap C_2$. Denote by $\tilde{F}_1$ and $\tilde{F}_2$ the outputs of the regularization algorithms with respect to C_1 and C_2. Both of them are more regular than $\bar{F}$ (since we have an optimization process and $\bar{F}$ is admissible), thus practically noise-free. The main difference between them should be the level of smoothing the image details, e.g., the contrast of the image edges should be smaller than the one for $\bar{F}$ and may vary between $\tilde{F}_1$ and $\tilde{F}_2$. Thus, taking the segmented difference image $|\tilde{F}_1 - \tilde{F}_2| > c$ with a proper threshold $c > 0$, should help us visualize and localize image regions, where important structural information is kept.

5 Numerical Results

We perform the proposed denoising methods on two different datasets of real-life radiographic images. The first one **Ex1** is a series of 16 images of size 723×920 pixels, 8-bit single-frame CT images of a 3D printed object. The second one **Ex2** is a series of 8 images of size 1446×1840 pixels, 16-bit single-frame CT images of a metal welding. For both datasets a 2048-frame-averaged image $avgF$ is available, which we use as an approximation of $\bar{F}$. Especially Ex2 shows the need of effective denoising algorithms as in this work piece there are small air bubbles included along the weld seam, possibly causing stability problems. In the original noisy image those bubbles are strongly overlaid by the noise which makes them almost invisible. Due to the large original size of the second example, we consider a cropped 500×600 version of it that covers the welding part.

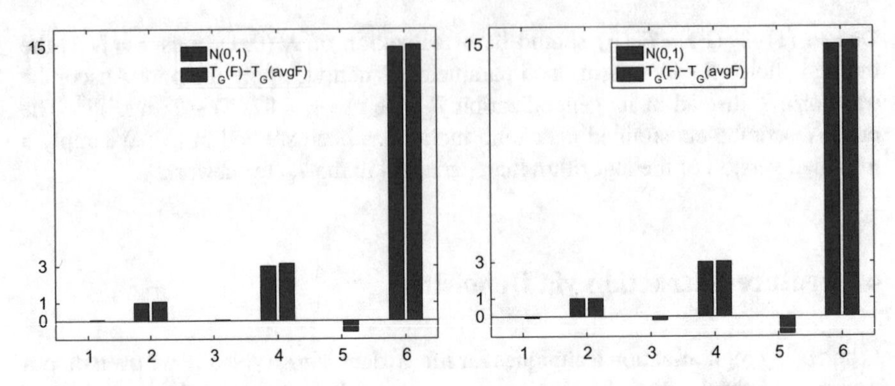

Fig. 2 Statistical analysis on noise characteristics. For each moment (*x-axis*), $N(0, 1)$ (*left bar* in each pair, *blue color*) is compared to $T_G(F) - T_G(avgF)$ (*right bar* in each pair, *red color*). Results for Ex1 are found in the *left figure* whilst results for Ex2 are found in the *right figure*

Applying the statistical analysis from Sect. 2.1 we estimate $(\alpha_1, \sigma_1^2) = (0.1708, 86.6697)$ for Ex1 and $(\alpha_2, \sigma_2^2) = (7.0961, 9523.12)$ for Ex2 in (1). As a numerical evidence for the reliability of the computed noise parameters, in Fig. 2 we compare the first six moments of $\{E((T_G(F) - T_G(avgF))^p)\}_{p=1}^6$ with those of $\mathcal{N}(0, 1)$, which are zero for all odd p and are equal to $(p-1)!! := 1 \cdot 3 \cdot 5 \dots (p-1)$ for even p. We observe that those theoretical values are very well approximated by the numerically computed ones, which is a strong indicator that $T_G(F) - T_G(avgF) \sim \mathcal{N}(0, 1)$. For large odd p we experience larger negative deviations from zero, which is related to the discussed earlier trimming of the distribution, due to the intensity range restriction $[0, \nu]$ (see Fig. 1).

As quantitative measurement of the denoising results we will use both the peak signal to noise ratio $\mathrm{PSNR}(\mathrm{avgF}, \hat{\mathrm{F}}) = 10 \log_{10} \frac{|\max \mathrm{avgF} - \min \mathrm{avgF}|^2}{\frac{1}{\mathrm{N}} \|\mathrm{avgF} - \hat{\mathrm{F}}\|_2^2}$ and the mean absolute error $\mathrm{MAE}(\mathrm{avgF}, \hat{\mathrm{F}}) = \frac{1}{\mathrm{N}\nu} \|\mathrm{avgF} - \hat{\mathrm{F}}\|_1$. Note that good denoising quality is indicated by high PSNR and low MAE.

Experimental results are summarized in Table 1 and visualized in Figs. 3 and 4. Only the continuous *L*2 model (6) gives rise to penalized optimization, where there is no statistical estimation of a good choice for λ. However, since the corresponding algorithm is direct and extremely fast, we run it for a large enough uniformly sampled discrete set of λ's and record the output with the highest PSNR (denoted by $\hat{F}_{L2,maxPSNR}$) and the output with the lowest MAE (denoted by $\hat{F}_{L2,minMAE}$). The other models (8), (9), (10) involve constrained optimization and the considered choices for the constraint parameter τ were already discussed in Sect. 3. Finally, we measure the TV-ratio $TV(avgF)/TV(\hat{F})$ as an indicator of the level of oversmoothing.

The first dataset Ex1 covers the full 8-bit intensity range. Due to the mean-variance relation for the Poisson distribution, noise is very well visualized on the light background (see Fig. 3) where the gray-scale intensities are close to 255. The variety in the sharpness of the image edges is also huge—we have high-contrast

Table 1 Quantitative results for the two data sets. For approximation of the true noise-free image $\bar{F}$ we use a 2048-frame-averaged image *avgF*

Image	3D printed object			Welding		
	PSNR	MAE (10^{-2})	TV-ratio	PSNR	MAE (10^{-2})	TV-ratio
F	26.5684	3.7023	6.2919	18.8539	3.6360	9.6743
$\hat{F}_{L2,minMAE}$	34.9932	1.1018	0.4919	32.8695	0.6650	0.1849
$\hat{F}_{L2,maxPSNR}$	35.5518	1.1578	0.6360	32.9135	0.6687	0.1951
$\hat{F}_{BM3D,T}$	31.2623	2.2428	0.3580	28.1649	1.3191	0.2041
$\hat{F}_{BM3D,T_G}$	33.0929	1.8195	0.3646	32.4835	0.7647	0.2371
$\hat{F}_{I-div}$	38.9435	0.8000	0.3465	34.5687	0.5946	0.1188
$\hat{F}_{TV,T}$	38.4642	0.7737	0.3241	32.5944	0.7447	0.1140
$\hat{F}_{TV,T_G}$	39.3334	0.7586	0.3463	32.5472	0.7406	0.1131

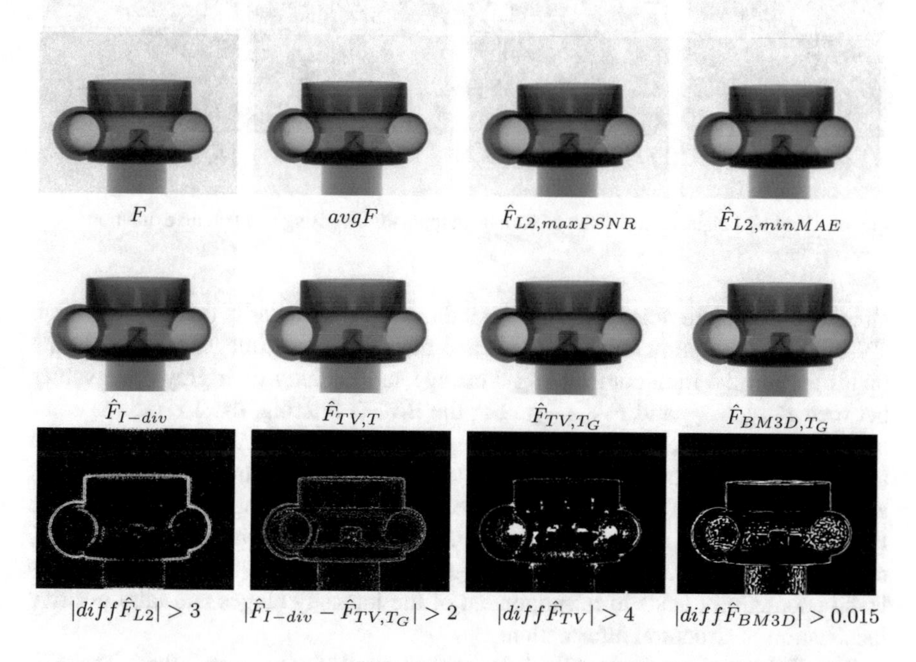

Fig. 3 3D printed object image. Comparison of various outputs for denoising and feature extraction

edges as well as low-contrast ones. Analyzing the numbers in Table 1 we observe that ℓ^2-regularization methods (6), (8) are clearly outperformed by the *TV*-regularization ones (9), (10). This is due to the fact that the mixed $\ell^{2,1}$ norm gives rise to sparser output gradients, so applying *TV*-regularization we first remove the noise, which is the least structured part of the image, and we start to smoothen the edges afterwards. This is not the case in ℓ^2-regularization, where high-contrast edges might be

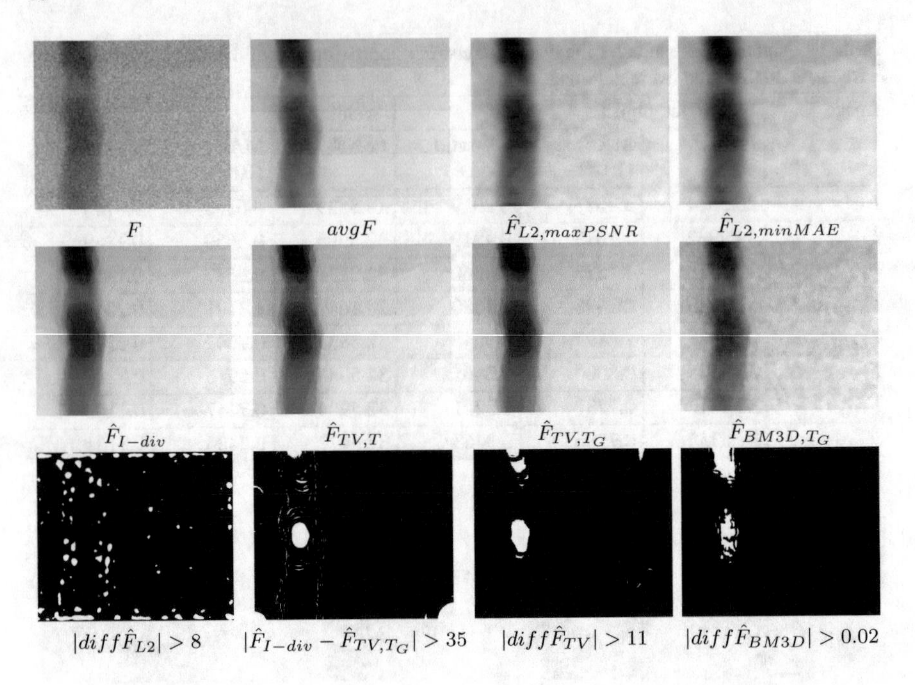

Fig. 4 Welding image. Comparison of various outputs for denoising and feature extraction

affected even before noise removal. An indicator for the latter is the relatively high TV-ratio for the optimal results of (6) and their difference image, where only the pixels around the high-contrast edges change substantially their gray-scale values between $\hat{F}_{L2,maxPSNR}$ and $\hat{F}_{L2,minMAE}$. For the BM3D filtering, the TV-ratio is comparable to the others, since a similar type of constrained optimization is performed and τ is carefully chosen, but the PSNR and MAE values are the worst. This is another confirmation that sharp edges are smoothed before the full noise removal. Here, there is also an artificial problem, since before and after the denoising process, current images are normalized to $[0, 1]$. As seen on their difference image, normalization sometimes leads to misalignment of the intensity ranges and false positive localization of structural information.

The TV-based methods (9), (10) behave similarly to each other. The T_G-constrained minimizer of (10) has better characteristics than the pure T-constrained one and preserves more structural information, as can be seen by their corresponding images and their difference image. It seems that optimal edge detection is achieved when comparing $\hat{F}_{TV,T_G}$ to $\hat{F}_{I-div}$.

Unlike Ex1, the second dataset Ex2 has a very narrow intensity range—all the gray-scale values of the pixels of *avgF* are in the interval $[2487, 4144]$, which length is less than $1/3$ of the admissible length 65535. There are no sharp edges and, as a result, all the four algorithms behave similarly. Since $\sigma_2 \gg 0$, the Gaussian noise component is not negligible and affects the quality of $\hat{F}_{BM3D,T}$. Indeed, we apply an AWGN filter to the non-Gaussian distribution $T(F) - T(avgF)$. Nevertheless the difference image $diff\hat{F}_{BM3D}$ can still be used for void detection, which is not true for the corresponding $diff\hat{F}_{L2}$, where, again, only the sharpest edges at the welding boundary are localized.

Knowing the correct noise parameters is not an advantage here for the quality of the T_G-constrained minimizer of (10). This information plays a role only for determining a meaningful τ, and if we can achieve the latter without the help of statistics (as is the case for the T-constrained minimizer, where $\tau = \|T_G(F) - T_G(avgF)\|_2^2$), the level of oversmoothing depends predominantly on the size of the constrained set. It is evident from the TV-ratio numbers, that the T_G-constrained set is larger, the result is more regular thus more structural information is lost, and the characteristics of the output are worse. This does not affect much the feature extraction procedure and, like for Ex1, the difference image $\hat{F}_{TV,T_G} - \hat{F}_{I-div}$ is the most useful one.

6 Conclusion

In this paper we experimentally compared various regularization-based denoising methods on real industrial CT radiographic data. Apart from noise removal, we managed to localize and extract important structural information about the scanned object only from the segmented difference image of denoised outputs. Which is the most suitable method for a given input image depends on the noise characteristics, that needs to be a priori examined, and on the level of contrast of the image features. However, I-divergence constrained TV-minimization (9) seems to be the most reliable one, provided the constrained parameter τ is optimally chosen. To the best of our knowledge, there is no theoretical result on a trustful estimation of τ for the case of mixed Poisson-Gaussian noise. On the other hand, the τ parameter of the Generalized Anscombe-constrained TV-minimization (10) is completely determined, once the noise is characterized, thus it seems a good practical choice. In any case, TV-regularization seems more robust with respect to noise removal than the corresponding ℓ^2-regularization techniques, and the difference image between the minimizers of (9) and (10) seems to capture the largest amount of structural information, thus it seems the best candidate for performing a feature extraction on.

Acknowledgements This work is a continuation of the research, performed during the *ECMI Modelling Week*, July 17–24, 2016, Sofia, Bulgaria. We are grateful to Ivan Georgiev (IICT-BAS) for providing us with real-life industrial CT data. The work of S. Harizanov has been partially supported by the "Program for career development of young scientists, BAS", grant No. DFNP-92/04.05.2016 and by the Bulgarian National Science Fund under grant No. BNSF-DM02/2 from 17.12.2016. The work of D. Wenzel has been supported by the *Institute of Numerical Analysis* of Dresden University of Technology.

References

1. Anscombe, F.J.: The transformation of poisson, binomial and negative-binomial data. Biometrika **35**(3/4), 246–254 (1948)
2. Azzari, L., Foi, A.: Variance stabilization for noisy+estimate combination in iterative poisson denoising. IEEE Signal Process. Lett. **23**, 1086–1090 (2016)
3. Harizanov, S., Pesquet, J.-C., Steidl, G.: Epigraphical projection for solving least squares Anscombe transformed constrained optimization problems. In: Scale-Space and Variational Methods in Computer Vision (SSVM 2013). LNCS, vol. 7893 pp. 125–136. Springer, Berlin (2013)
4. Teuber, T., Steidl, G., Chan, R.H.: Minimization and parameter estimation for seminorm regularization models with I-divergence constraints. Inverse Prob. **29**, 1–28 (2013)
5. Starck, J.-L., Murtagh, F., Bijaoui, A.: Image Processing and Data Analysis: The Multiscale Approach. Cambridge University Press, New York (1998)
6. Mäkitalo, M., Foi, A.: Optimal inversion of the generalized anscombe transformation for poisson-gaussian noise. IEEE Trans. Image Process. **22**, 91–103 (2013)
7. Dabov, K., Foi, A., Katkovnik, V., Egiazarian, K.: Image denoising by sparse 3-D transform-domain collaborative filtering. IEEE Trans. Image Process. **16**, 2080–2095 (2007)
8. Foi, A., Katkovnik, V., Egiazarian, K.: Pointwise shape-adaptive DCT for high-quality denoising and deblocking of grayscale and color images. IEEE Trans. Image Process. **16**, 1395–1411 (2007)
9. Buades, A., Coll, B., Morel, J.M.: A review of image denoising algorithms, with a new one. Multiscale Model. Simul. **4**(2), 490–530 (2005)

Representation of Civilians and Police Officers by Generalized Nets for Describing Software Agents in the Case of Protest

Shpend Ismaili and Stefka Fidanova

Abstract Agent-based modeling and simulation to solve difficult problems, becomes very popular last years. Predicting and preventing conflict situations are very actual now days. Therefore various mathematical techniques are used. One of them is application of multi-agent systems. The main element of the multi-agent systems is the software agent, which is an autonomous subject with a possibility to work together with other agents and environment. In our application the software agents represent civilians and police officers in protests. In this work we propose a model of the software agents with Generalized Nets. The Generalized Net is a very powerful tool for modeling processes and different situations. They are expandable and can represent a process in details. In this work we propose a model of the software agents with Generalized Net. Our agents model the behavior of the civilian and police officers in case of the protest.

1 Introduction

During some protest very important is the possibilities for crowd control and preventing and elimination of conflict situations. Very often there are factors which are difficult to predict, even when we expect that the crowd is well managed. In this case can appear conflict situations. It can cause mess and casualties. Crowd simulation is a very important research topic. Various approaches are applied for researching crowd behavior, fuzzy-theory-based method [9], bandit strategy [5], cellular automata [13], crowd motion simulation [12].

One of the simulation methods which is applied on modeling crowd behavior is agent-based. Multi-agent system consists of different kind of agents and environ-

S. Ismaili
University of Tetovo, Tetovo, Macedonia
e-mail: shpend.ismaili@unite.edu.mk

S. Fidanova (✉)
Institute of Information and Communication Technology,
Bulgarian Academy of Science, Sofia, Bulgaria
e-mail: stefka@parallel.bas.bg

K. Georgiev et al. (eds.), *Advanced Computing in Industrial Mathematics*, Studies in Computational Intelligence 728,
https://doi.org/10.1007/978-3-319-65530-7_7

ments. The interaction between the agents and change of the environment affect the individual agent and it can change his behavior. The agents can be passive or active and can react in different manner according the situation [18].

Generalized Nets (GN) [1–3] are an efficient tool for modeling of various real processes. They are extension of Petri nets. The apparatus of the GN is very powerful and can be used for modeling in different areas like medicine and biology, economics, industry, description of algorithms and many others [16, 19–21].

In this paper GN are used as a tool for modeling of software agents in multi agent system with application in simulation of conflicting situations in the case of protest.

The rest of the paper is organized as follows. In Sect. 2, we give short description of the main elements from GN-theory. In Sect. 3 the problem is defined. In Sect. 4 the software agents are describe with a GN. At the end we give some conclusions.

2 Short Description of the GN

The GN was proposed for a first time in 1991 [2]. Later they was applied for description of different processes and algorithms [8, 11, 15, 17]. They are powerful tool for description of complex systems with not homogeneous components. Its static structure consists of objects called **transitions**, which have input and output **places**. Two transitions can share a place, but every place can be an input of at most one transition and can be an output of at most one transition.

The dynamic structure consists of **tokens**, which act as information carriers and can occupy a single place at every moment of the GN execution. The tokens pass through the transition from one input to another output place; such an ordered pair of places is called **transition arc**. The tokens' movement is governed by conditions (predicates), contained in the **predicate** matrix of the transition.

The information carried by a token is contained in its **characteristics**, which can be viewed as an associative array of characteristic names and values. The values of the token characteristics change in time according to specific rules, called **characteristic functions**. Every place possesses at most one characteristic function, which assigns new characteristics to the incoming tokens. Apart from movement in the net and change of the characteristics, tokens can also split and merge in the places. A transition can contain m input and n output places where $n, m \geq 1$.

The GN can be expanded. The places can be replaced with other GN. In this case the GN can be developed in steps, including new details. Thus we can see possibilities for other development and can better understand the processes.

Formally, every transition is described by a seven-tuple (Fig. 1):

$$Z = \langle L', L'', t_1, t_2, r, M, \square \rangle,$$

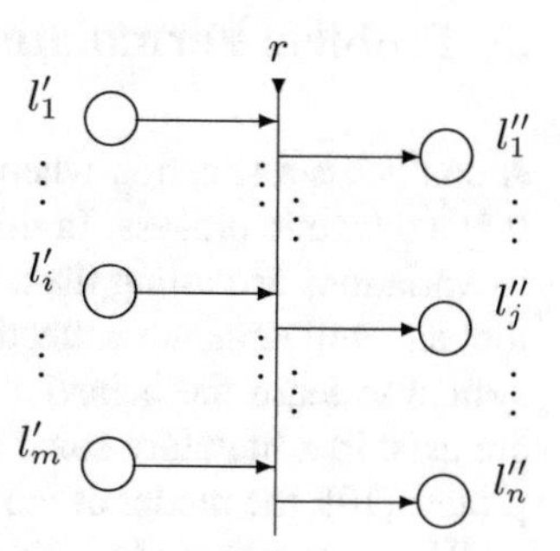

Fig. 1 The form of one transition

where:

(a) L' and L'' are finite, non-empty sets of places (the transition's input and output places, respectively); for the transition in Fig. 1 these are $L' = \{l'_1, l'_2, \ldots, l'_m\}$ and $L'' = \{l''_1, l''_2, \ldots, l''_n\}$;
(b) t_1 is the current time-moment of the transition's firing;
(c) t_2 is the current value of the duration of its active state;
(d) r is the transition's *condition* determining which tokens will pass (or *transfer*) from the transition's inputs to its outputs; it has the form of an Index Matrix (IM; see [4]):

$$r = \begin{array}{c|ccccc} & l''_1 & \ldots & l''_j & \ldots & l''_n \\ \hline l'_1 & & & & & \\ \vdots & & & r_{i,j} & & \\ l'_m & & & & & \end{array};$$

$r_{i,j}$ is the predicate that corresponds to the i-th input and j-th output place ($1 \leq i \leq m, 1 \leq j \leq n$). When its truth value is "*true*", a token from the i-th input place transfers to the j-th output place; otherwise, this is not possible;
(e) M is an IM of the capacities $m_{i,j}$ of transition's arcs, where $m_{i,j} \geq 0$ is a natural number:

$$M = \begin{array}{c|ccccc} & l''_1 & \ldots & l''_j & \ldots & l''_n \\ \hline l'_1 & & & & & \\ \vdots & & & m_{i,j} & & \\ l'_m & & & & & \end{array};$$

(f) $\square$ is the transition type, it is an object of a form similar to a Boolean expression. It contains as variables the symbols that serve as labels for a transition's input places, and $\square$ is an expression built up from variables and the Boolean connectives $\wedge$ and $\vee$. When the value of a type (calculated as a Boolean expression) is "*true*", the transition can become active, otherwise it cannot.

3 Problem Formulation

A conflict is a situation where minimum two persons, strive to achieve their goals. It is a dynamic process. In this paper we try to understand the human behavior and its variations according the situation. We try to represent different groups in a conflict and will simulate collective behavior. We focus on development of multi agent system to learn the behavior caused by the interaction between the agents. There are exist in a literature some computer models of concrete protests: model of trade protest [10]; the model of violence in London [6]; model of revolution [14].

We create more individuals which interact between them, to model civil violence. The structure consists of individuals, environment and empirical rules. Our software agents model polis officers and civilians. Accurate modeling of their attributes is crucial to the description which is as much as possible closer to human life and behavior in situations of unrest. Peaceful civilians are neutral participant, but they can react to external or internal stimulus. Police officers retain the order by the insertion of the activists in jail and through strategies that choose depends on the success of the management and control of violence. The police officers perform two tasks in a direct way: active arrest protesters and move in space.

Civilians are much more complex individuals, than the police officers. Civilian agent decides whether to be active or not. Typical of civilian agents is communication. The civilian agents can change from active to passive and from passive to active. The functioning of the system depends of the empirical rules. Empirical rules guide the interactions of agents and ensure the functioning of the system.

4 GN for Software Agents

In this section we propose representation logic of civilians agents and police agents by GN in the case of protest. First we establish the status of civilian.

- Status of civilian, prisoner or free;
- If he is prisoner, whether he served a term of prison or not;
- If he is free, continue to move and take independent decisions whether to participate in the protest or not.

We will introduce the terms active and peaceful and level of discontent (NAI) and threshold (Athreshold) for danger [7], where NAI = Rev – N; Rev-tendency to revolt and N is a net risk (the risk of imprisonment).

According the relation between *NAI* and *Athreshold*, it will be the following cases:

- If the civilian is peaceful and $NAI > Athreshold$, he will become active;
- If the civilian is active and $NAI > Athreshold$, he will stay active;
- If the civilian is peaceful and $NAI < Athreshold$, he will stay peaceful;
- If the civilian is active and $NAI < Athreshold$, he will become peaceful.

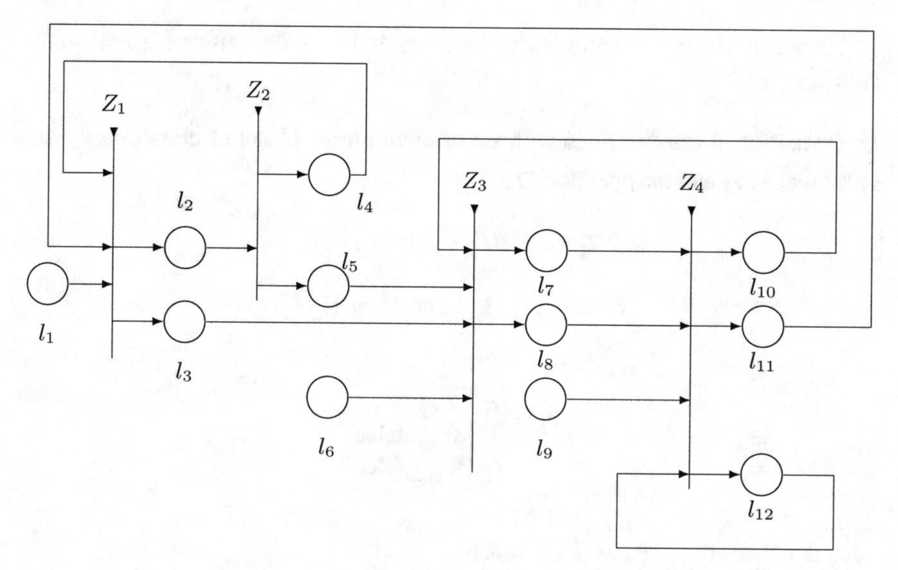

Fig. 2 GN for agents representation in the case of protest

After that is activated the police officers logic:

- If the police officer encounters peaceful civilian, he track him in his zone of monitoring;
- If the police officer encounter active civilian, he closes him;
- If there are not persons in the police officer zone of monitoring, he moves in the random way.

We use GN with 4 transitions (Z_1, Z_2, Z_3, Z_4) and 12 places ($l_1, \ldots, l_{12}$) to represent different kind of agents.

The meaning of the indications on the Fig. 2 are:

l_1 – civilian without a certain position
l_2 – civilian is detained
l_3 – civilian is free
l_4 – the term of imprisonment has not expired
l_5 – the term of imprisonment has expired
l_6 – active civilian
l_7 – will be peaceful civilian
l_8 – will be active civilian
l_9 – police officer
l_{10} – peaceful civilian
l_{11} – prisoner
l_{12} – random movement of the police in his area of monitoring

There are four transition in our GN representation of the software agents in the case of protest:

Transition Z_1: is described with set of input places L', set of output places L'', index matrix r_1 and the operator $\Box$.

$$Z_1 = \langle L', L'', r_1, \vee(l_1, l_4, l_{11})\rangle,$$

$$L' = \{l_1, l_4, , l_{11}\} \text{ and } L'' = \{l_2, l_3\}$$

$$r_1 = \begin{array}{c|cc} & l_2 & l_3 \\ \hline l_1 & W_{1,2} & W_{1,3} \\ l_4 & W_{4,3} & \text{falce} \\ l_{11} & W_{11,2} & \text{falce} \end{array};$$

where
$W_{1,2} = W_{4,2} = W_{11,2}$ there is a prisoner
$W_{1,3}$ civilian is free

Transition Z_2: Statute of prisoner.

$$Z_2 = \langle \{l_2\}, \{l_4, l_5\}, r_2, \vee(l_4, l_5)\rangle,$$

$$r_2 = \begin{array}{c|cc} & l_4 & l_5 \\ \hline l_2 & W_{2,4} & W_{2,5} \end{array};$$

where
$W_{2,4}$ the term of imprisonment has not expired
$W_{2,5}$ the term of imprisonment has expired

Transition Z_3: Statute of civilians.

$$Z_3 = \langle \{l_3, l_5, l_6, l_{10}\}, \{l_7, l_8\}, r_2, \vee(l_7, l_8)\rangle,$$

$$r_3 = \begin{array}{c|cc} & l_7 & l_8 \\ \hline l_3 & W_{3,7} & W_{3,8} \\ l_5 & W_{5,7} & W_{5,8} \\ l_6 & W_{6,7} & W_{6,8} \\ l_{10} & W_{10,7} & W_{10,8} \end{array};$$

where
$W_{3,7} = W_{5,7}$ $NAI < Athreshold$ => peaceful
$W_{3,8} = W_{5,8}$ $NAI > Athreshold$ => active
$W_{6,7}$ $NAI < Athreshold$ => peaceful
$W_{6,8}$ $NAI > Athreshold$ => active
$W_{10,7}$ $NAI < Athreshold$ => peaceful
$W_{10,8}$ $NAI > Athreshold$ => active

Transition Z_4: Police officer logic.

$$Z_4 = \langle \{l_7, l_8, l_9, l_{12}\}, \{l_{10}, l_{11}, l_{12}\}, r_2, \vee(l_{10}, l_{11}, l_{12}) \rangle,$$

$$r_4 = \begin{array}{c|ccc} & l_{10} & l_{11} & l_{12} \\ \hline l_7 & true & false & false \\ l_8 & false & W_{8,11} & false \\ l_9 & false & W_{9,11} & true \\ l_{12} & false & W_{12,11} & W_{12,12} \end{array};$$

where
$W_{8,11} = W_{9,11} = W_{12,11}$ the agent to be arested/stay in the prison
$W_{10,8}$ random movement

5 Conclusion

The constructed model can be used for simulation of the behavior of the participants of a case of protest. It can be expanded including more possibilities. By this kind of models various situations can be play in advance and can be predicted and prevented serious conflicts. Including more details some specific protests can be modeled and the acts of the police officers can be decided and trained before the protest.

Acknowledgements Work presented here is partially supported by the Bulgarian National Scientific Fund under the grants DFNI-I02/20 “Efficient Parallel Algorithms for Large Scale Computational Problems” and DFNI DN 02/10 “New Instruments for Data Mining and their Modeling”.

References

1. Alexieva, J., Choy, E., Koycheva, E.: Review and bibloigraphy on generalized nets theory and applications. In: Choy, E., Krawczak, M., Shannon, A., Szmidt, E. (eds.) A Survey of Generalized Nets, Raffles KvB Monograph No. 10, pp. 207–301 (2007)
2. Atanassov, K.: Generalized Nets. World Scientific, Singapore, London (1991)
3. Atanassov, K.: On Generalized Nets Theory. Prof. M. Drinov Academic Publishing House, Sofia (2007)

4. Atanassov, K.: Index Matrices: Towards an Augmented Matrix Calculus. Springer, Cham (2014)
5. Chen, H., Rahwan, I., Cebrian, M.: Bandit strategies in social search: the case of the DARPA red balloon challenge. EPJ Data Sci. **5**(1), 20 (2016)
6. Davies, T.P., Fry, H.M., Wilson, A.G., Bishop, S.R.: A mathematical model of the London riots and their policing. Sci. Rep. **3**, 1303 (2013)
7. Epstein, J.M.: Modeling civil violence: an agent-based computational approach. In: Proceedings of the National Academy of Sciences of the United States of America, vol. 99, pp. 7243–7250 (2002)
8. Fidanova, S., Atanassov, K., Marinov, P.: Generalized Nets and Ant Colony Optimization. Academy of Sciences Publishing House, Bulg (2011). ISBN 978-954-322-473-9
9. Fu, L., Song, W., Lo, S.: A fuzzy-theory-based method for studying the effect of information transmission on nonlinear crowd dispersion dynamics. Commun. Nonlinear Sci. Numer. Simul. **42**, 682–698 (2017)
10. Kim, J.W., Hanneman, R.A.: A computational model of worker protest. J. Artif. Soc. Soc. Simul. **14**(3), (2011)
11. Krawczak, M.: A Novel modeling methodology: generalized nets. In: Artificial Intelligence and Soft Computing. LNCS, vol. 4029, pp. 1160–1168. Springer (2006)
12. Li, D., Yuan, L., Hu, Y., Zhang, X.: Large-scale crowd motion simulation based on potential energy field. J. Huazhong Univ. Sci. Technol. **44**(6), 117–122 (2016)
13. Lubas, R., Was, J., Porzycki, J.: Cellular Automata as the basis of effective and realistic agent-based models of crowd behavior. J. Supercomput. **72**(6), 2170–2196 (2016)
14. Makowsky, M.D., Rubin, J.: An agent-based model of centralized institutions, social network technology, and revolution. Working paper 2011–05, Towson University Department of Economics (2011)
15. Peneva, D., Tasseva, V., Kodogiannis, v., Sotirova, E.: Generalized nets as an instrument for description of the process of expert system construction. In: IEEE Intelligent Systems, pp. 755–759 (2006)
16. Ribagin, S., Roeva, O., Pencheva, T.: Generalized net model of asymptomatic osteoporosis diagnosing. In: 2016 IEEE 8th International Conference on Intelligent Systems (IS 2016), pp. 604–608, 7 Nov 2016
17. Shannon, A., Sorsich, J., Atanassov, K.: Generalized Nets in Medicine. Prof. M. Drinov Academic Publishing House, Sofia (1996)
18. Salamon, T.: Design of Agent-Based Models: Developing Computer Simulations for a Better Understanding of Social Processes. Bruckner Publishing (2011). ISBN 978-80-904661-1-1
19. Sotirov, S., Sotirova, E., Werner, M., Simeonov, S., Hardt, W., Simeonova, N.: Ituitionistic fuzzy estimation of the generalized nets model of spatial-temporal group scheduling problems. Studies in Fuzziness and Soft Computing, vol. 332, pp. 401–414. Springer (2016)
20. Sotirova, E., Bureva, V., Sotirov, S.: A generalized net model for students evaluation process using intercriteria analysis method. Studies in Fuzziness and Soft Computing, vol. 332, pp. 389–399. Springer (2016)
21. Stefanova-Pavlova, M., Andonov, V., Stoyanov, T., Angelova, M., Cook, G., Klein, B., Vassilev, P., Stefanova, E.: Modeling telehealth services with generalized nets. Stud. Comput. Intell. **657**, 279–290 (2017)

Comparison of NDT Techniques for Elastic Modulus Determination of Laminated Composites

Yonka Ivanova, Todor Partalin and Ivan Georgiev

Abstract The study of dependence of elasticity modulus on the type, shape and structure of the fillers in the composites is an important task. By theoretical point of view, different models are developed to describe the relation between geometrical, mechanical and physical parameters of the fillers and matrix with the macroscopic effective properties of composites. It is reasonably the Young's modulus of the composites to be determined both experimentally and theoretically.In the present study non-destructive techniques are used for characterization of elastic modulus. NDT methods, static and dynamic ultrasonic and vibration methods are applied to find the relations between internal structure of composites and their elastic properties.For investigating of the elastic properties in different directions of composites are used a methodology based on combination of different kind of vibrations. The bar shaped specimens are examined by free longitudinal, flexural and torsional vibrations.

Keywords Laminated composites · Elastic modulus · Non-destructive techniques

Y. Ivanova
Faculty of Physics, Sofia University "St.Kliment Ohridski", Sofia, Bulgaria
e-mail: yonka@imbm.bas.bg

Y. Ivanova
Institute of Mechanics, Bulgarian Academy of Sciences, Sofia, Bulgaria

T. Partalin
Faculty of Mathematics and Informatics, Sofia University "St.Kliment Ohridski", Sofia, Bulgaria
e-mail: topart@fmi.uni-sofia.bg

I. Georgiev (✉)
Institute of Information and Communication Technologies, Bulgarian Academy of Sciences, Sofia, Bulgaria
e-mail: ivan.georgiev@parallel.bas.bg

I. Georgiev
Institute of Mathematics and Informatics, Bulgarian Academy of Sciences, Sofia, Bulgaria

K. Georgiev et al. (eds.), *Advanced Computing in Industrial Mathematics*, Studies in Computational Intelligence 728,
https://doi.org/10.1007/978-3-319-65530-7_8

1 Introduction

In this paper the engineering constants of laminated composites plates are determined by non-destructive methods and techniques based on static and dynamic approaches. A glass fabric laminated composite with thickness of 4 mm (GFC4) was chosen for the study. The plate was manufactured by hot pressing the glass cloth layers, impregnated with thermo-reactive phenolic and epoxy type resins (Electra LTD, Ruse, Bulgaria).

The glass fiber volume fraction was obtained as 45%. The density was determined by measuring the mass and computing the volume (mass was measured with high precision electronic balance, volume was computed by measuring the dimensions with a digital caliper). The material anisotropy was investigated by cutting the samples upon 0°, 30°, 45° and 90° on the length. The scheme of preparation of the samples is shown in Fig. 1, where the axis x is oriented collinear to weft and the axis z is oriented collinear to warp of the fabric. The geometry, stacking sequence, and density of each test specimen are given in Table 1.

2 Non-destructive Evaluation of Elastic Properties: Static Approach

A non-destructive static approach based on the four-point-bending test principle [1–3] is applied using beam like samples in order to determine their elastic properties (Fig. 2). A simply supported beam is loaded with two equal and equidistant point forces P on either side of the two rollers. In the middle span between the two supports the bending moment M is constant, i.e. middle span has "pure" bending. The beam segment bends in the shape of a circular arc of radius "ρ". The curvature $k = \frac{1}{\rho}$ is related to the bending moment M with the relation [1]:

$$k = \frac{1}{\rho} = \frac{M}{EI} \tag{1}$$

where E is the Young's modulus and I is the area moment of inertia of the cross-section of the beam. Since the beam has a rectangular cross section of width "b" and thickness "h", the area moment of inertia I is $I = bh^3/12$ [1].

In a four point bending configuration, the magnitude of the constant bending moment M in middle span is:

$$M = P \cdot a, \tag{2}$$

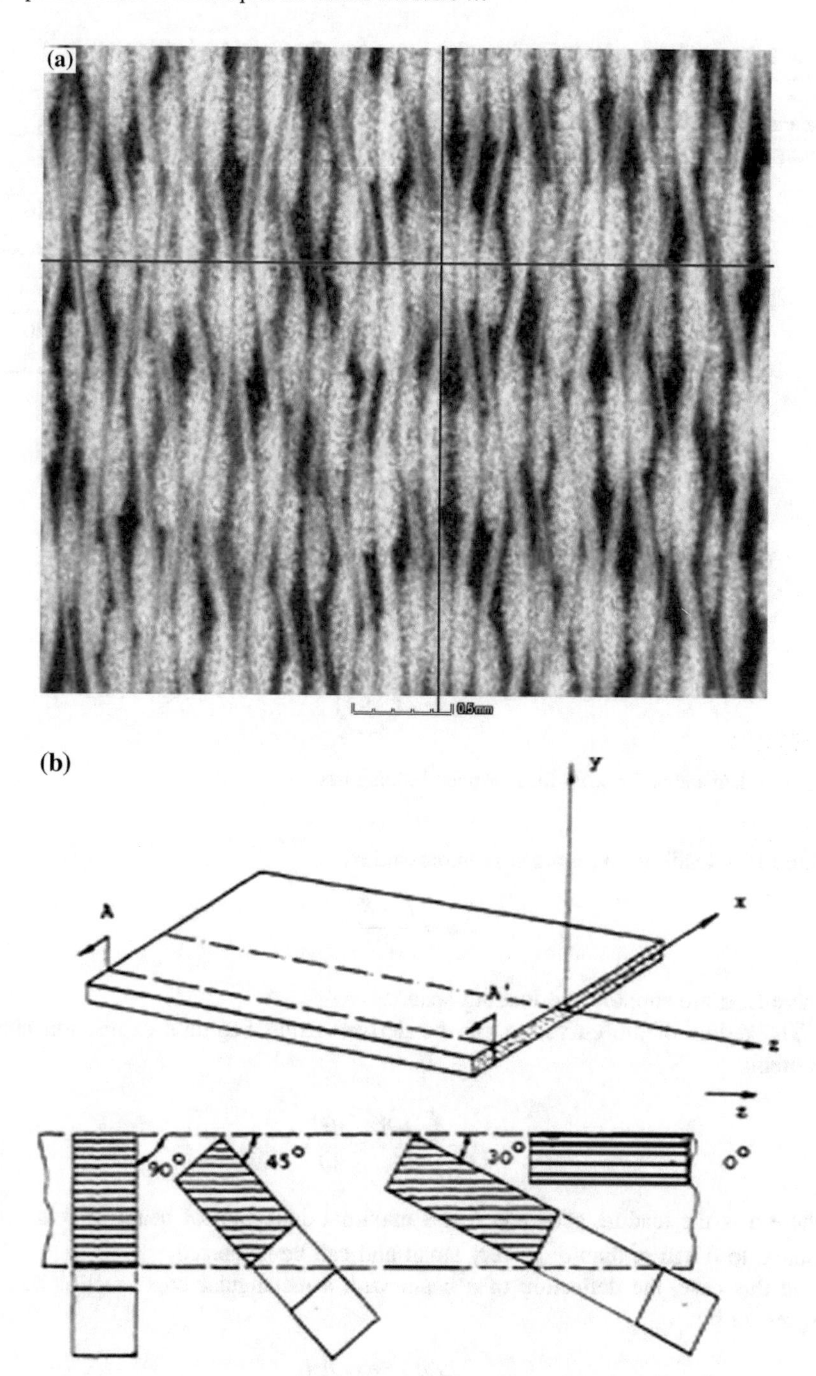

Fig. 1 Glass fabric composite: **a** micro CT image of the structure of the composite, **b** specimens preparation

Table 1 Type of materials, density, geometry, orientation

Type material	Glass content (%)	Density (kg/m^3)	Specimens	Stacking sequence	Thickness (m)	Width (m)	Length (m)
GFC4	45	1820	Beam 1,2,3	0°	0.00378	0.02	0.15
			Beam 4,5,6	30°	0.00378	0.02	0.16
			Beam 7,8,9	45°	0.00378	0.02	0.12
			Beam 10,11,12	90°	0.00378	0.02	0.10

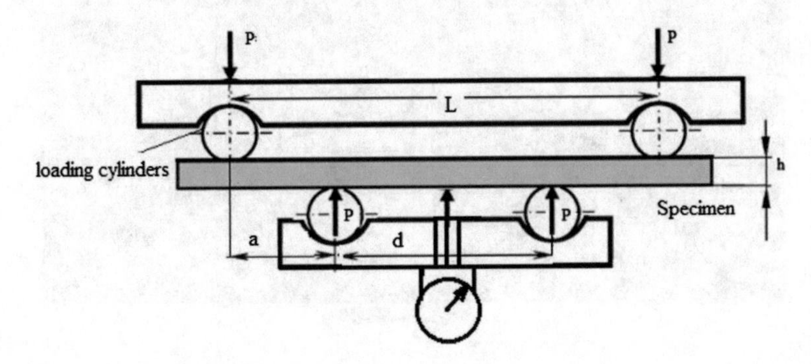

Fig. 2 Schematic of the setup for four-point bending test

where P is loading force and a is expressed as:

$$a = \frac{L - d}{2}. \tag{3}$$

where L, d are support and loading span.

The radius of the curvature can be derived using Sagitta's expression chord theorem:

$$\rho = \frac{\frac{d^2}{4} + \delta^2}{2\delta} \approx \frac{d^2}{8\delta} \tag{4}$$

where d is the loading span and δ is a maximal deflection of beam. If δ is small relative to d and ρ, then δ^2 is very small and can be neglected.

In this case, the deflection of a beam with a rectangular cross section can be expressed by:

$$\delta = \frac{Md^2}{8EI} = \frac{3Md^2}{2bh^3}\frac{1}{E} \tag{5}$$

The elasticity modulus is given by Eq. 6

$$E = \frac{M\rho}{I} = \frac{3Pd^2(L-d)}{4\delta bh^3} \quad (6)$$

The experimental setup consists of two roller supports for a rectangular beam, two weight hangers located at distance "a" from the load supports, and digital indicator, which is placed in the middle of the beam length and measures the maximal beam deflection.

Figure 3 presents dependencies of bending moment M_{max} and deflection of the beams. Using Eq. (6), the Young's moduli of each beam have been calculated and listed in Table 2.

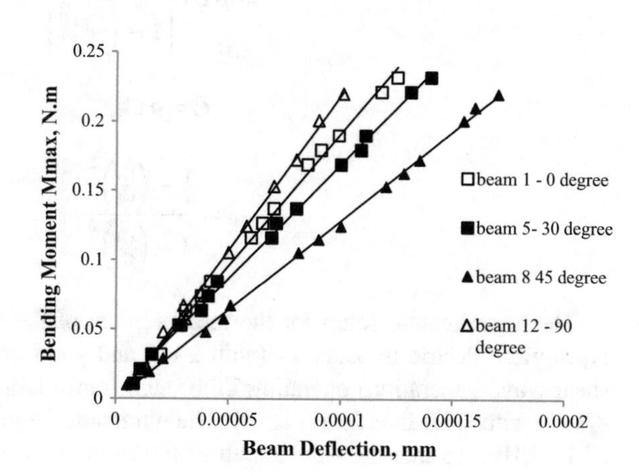

Fig. 3 Bending moment—beam deflection

Table 2 Elasticity modulus E (GPa)

Orientation	Static method		Dynamic methods						
			Flexural mode		Longitudinal mode		Ultrasonic method		
	$\overline{E}$	S	$\overline{E}$	S	$\overline{E}$	S	$\overline{E}$	S	Poisson ratio (ν)
0° Beams 1,2,3	21.5	0.215	21.4	1.61	22.4	1.51	13.5	0.347	0.33
30° Beams 4,5,6	16.3	0.997	16.3	0.475	16.4	0.593			
45° Beams 7,8,9	13.2	0.743	15.1	0.652	15.4	0.581			
90° Beams 10,11,12	21.2	0.751	21.3	1.96	23.0	3.67			

3 Non-destructive Evaluation of Elastic Properties: Dynamic Methods

Dynamic methods provide an advantage over static methods because of wide variety of specimen shapes and sizes and great precision [4–8]. They are classified into impulse (ultrasonic) and resonance methods.

Determination of the dynamic elastic moduli by ultrasound is based on the relation between ultrasonic longitudinal (C_L) and shear waves (C_S), propagating in materials and density (ρ), Young's modulus (E), Shear modulus (G) and Poisson's ratio ν [4]:

$$E = 4\rho\, C_S^2 \left[\frac{\frac{3}{4} - \left(\frac{C_S}{C_L}\right)^2}{1 - \left(\frac{C_S}{C_L}\right)^2} \right] \tag{7}$$

$$G = \rho\, C_S^2 \tag{8}$$

$$\nu = \frac{\frac{1}{2} - \left(\frac{C_S}{C_L}\right)^2}{1 - \left(\frac{C_S}{C_L}\right)^2} \tag{9}$$

The experimental setup for the measurement of the velocities consisted of two types piezoelectric transducers (with x-cut and y-cut crystals for longitudinal and shear wave generation) operating in through transmission mode [4] and ultrasonic system with USB Interface (Fig. 4). The ultrasonic frequency used was in the range of 1.5 MHz, so that the wavelength of ultrasonic waves was much larger than the glass fiber diameters.

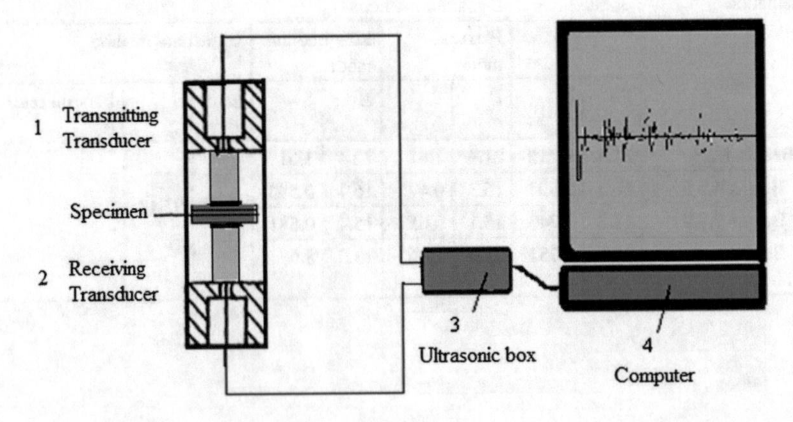

Fig. 4 Scheme of ultrasonic measurement

Table 3 Shear modulus G (GPa)

Orientation	Dynamic vibration torsional mode		Dynamic ultrasonic method	
	G	S	G	S
0° Beams 1,2,3	5.36	0.147	5.09	2.53
30° Beams 4,5,6	8.55	0.0763		
45° Beams 7,8,9	9.52	0.0246		
90° Beams 10,11,12	5.63	0.230		

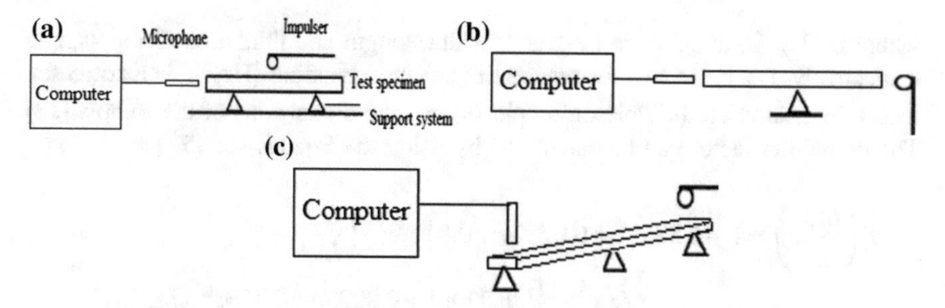

Fig. 5 Schematic of support setup for impulse excitation technique: **a** flexural mode, **b** longitudinal mode, **c** torsional mode of vibration

Measured values of longitudinal and shear wave velocities are used in Eqs. (7)–(9) to calculate the elastic (E) and shear (G) moduli in transversal direction of fiber-glass plates. The obtained values of moduli refer to the whole material and are listed in the Tables 2 and 3.

The resonance vibration method and techniques are well established and widely used techniques for the determination of the dynamic elastic properties of a large diversity of materials (glass, ceramic, concrete, composites, steels, etc.). They are covered by several ASTM standards [5–7]. The techniques consist in exciting a vibration by drivers having continuously variable frequencies output or by impact [5–14].

Knowing the vibrational mode, frequency, dimensions and mass or density of the samples it is possible to calculate the effective elastic modulus of the materials [5–7, 9, 10] by substituting in the appropriate frequency equation which is derived from the equation of motion for the specimen [9, 10]. The use of an effective modulus in composites is based on the assumption that the wavelength associated with the particular vibrational mode is much greater than the scale of the inhomogeneity in the composite.

In free vibration method the impulse excitation is produced by striking the object with a suitable hammer. As a pickup transducer is used acoustic microphone which signal is addressed to personal computer with a sound card and processed by signal processing methods (Fourier transform algorithm) in order to identify the values of the natural frequencies of vibration.

The schematic depiction of experimental set-up for flexural mode of vibration is shown in Fig. 5a. The sample was put onto two fulcrums at 0.224 times the total length from the ends of the sample [5, 6, 13, 16]. To induce a vibration in the rectangular shaped specimens, a rubber hammer is used to impact at the center of the sample. The sound produced by the vibration of the specimen is detected by means of a microphone and processed to make modal frequencies analyses. The Young's modulus of a beam is calculated following [5, 6, 8, 13, 16]:

$$E = 0.9465\rho \cdot f_{f,1}^2 \cdot L_x^2 \left(\frac{L_x^2}{L_z^2}\right) T\left(\frac{L_z}{L_x}, \nu\right) \tag{10}$$

where m, L_y, L_x and L_z are the mass, width, length and thickness of the sample, respectively, $f_{f,1}$ is the first resonance frequency in bending (Hz), T is a correction factor depending on the Poisson's ratio (ν) and the dimensions of the sample [13]. The correction factor can be calculated by using the Eq. 11, see [5, 13, 15–17].

$$\begin{aligned} T\left(\frac{L_z}{L_x}, \nu\right) &= 1 + 6.585\left(1 + 0.0752\nu + 0.8109\nu^2\right)\left(\frac{Lz}{Lx}\right)^2 \\ &\quad - 0.868\left(\frac{Lz}{Lx}\right)^4 - \left[\frac{8.340(1 + 0.2023\nu + 2.173\nu^2)(Lz/Lx)^4}{1 + 6.338(1 + 0.1408\nu + 1.536\nu^2)(Lz/Lx)^2}\right] \end{aligned} \tag{11}$$

The numerical analysis [16] shows that the error of formulas (11) is smaller than 1% when the Poisson ratio is smaller than 0.35, and the length-to-width ratio of the sample is larger than about 2.

The setup for free longitudinal vibration test is shown in Fig. 6b. Each specimen is hold from its center and is hit by a plastic hammer at the end of specimen. To analyze the acoustic response of the specimen, a microphone is positioned in the other side of sample. The recorded signals are analyzed by the means of Fourier Transform. The standing waves are formed in the specimens as the results of longitudinal vibration of the sample. Generally, sound velocity in a specimen could be determined from Eq. 12 [13–15, 18]:

$$\lambda = C/f_l \tag{12}$$

where C is sound velocity in a specimen, and f_l is the first mode of vibration resonance frequencies. The wave length λ is calculated from Eq. 13:

$$\lambda = 2L_x/n \tag{13}$$

where L_x is the length of specimen and n is the number of resonance mode. According to the positions of the node and two antinodes corresponding to the first mode of vibration, the wave length is equal to the twice of the specimen length.

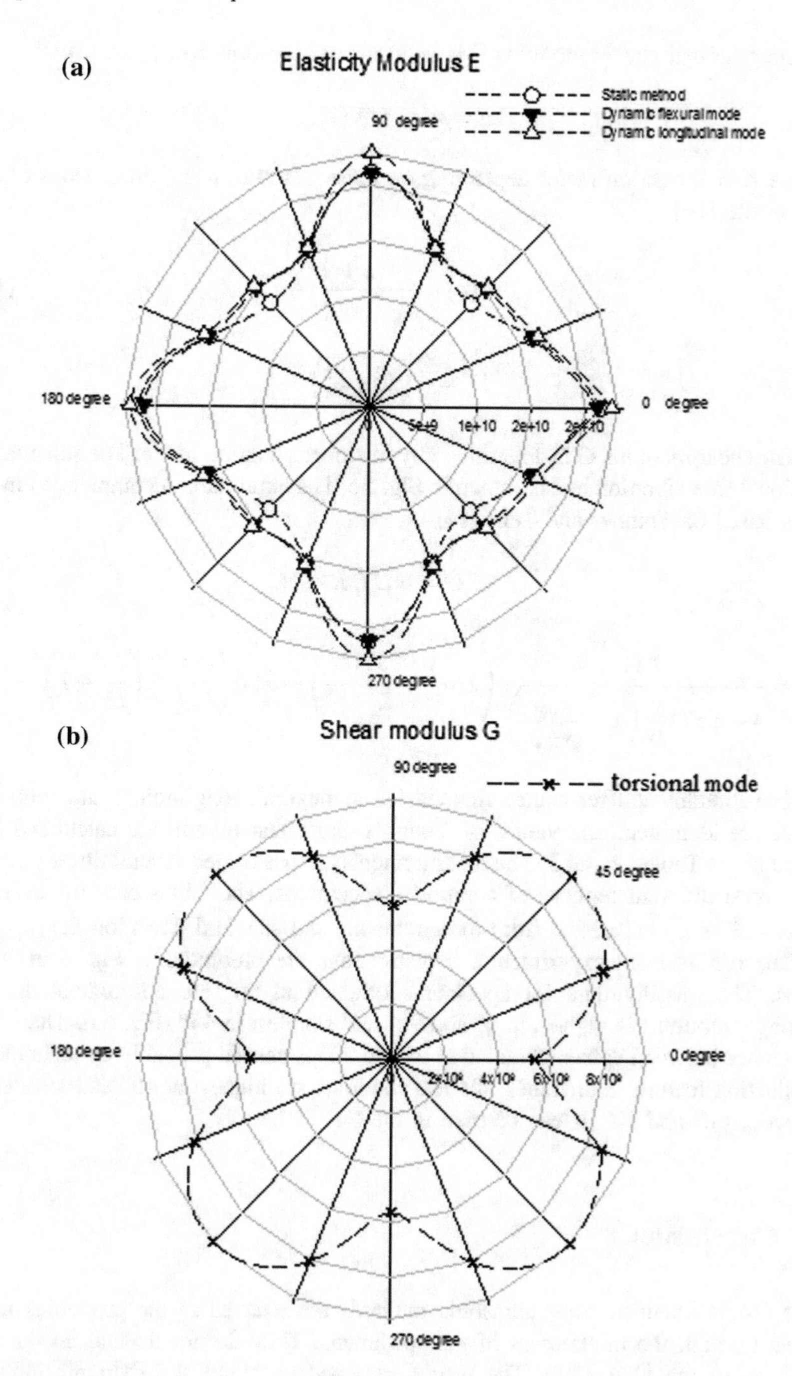

Fig. 6 Orientation of determined Young's (**a**) and Shear moduli (**b**)

Longitudinal elastic modulus E is determined according to Eq. 14 [6]:

$$E = \rho C^2 \cdot 1/K(\nu), \tag{14}$$

where K is correction factor depending on Poisson's ratio ν and dimensions of the composite [16]

$$K(\nu) = 1 - \frac{\pi^2 \nu^2 D^2}{8L_x^2} \tag{15}$$

$$D^2 = \frac{2}{3}\left(L_y^2 + L_z^2\right) \tag{16}$$

The shear modulus G is determined by torsional vibration mode. The schema for torsional free vibration test is shown in Fig. 5c. The expression recommended in [5, 6] is based on Spinner and Tefft [14]

$$G = 4\rho L_x^2 f_t^2 R \tag{17}$$

$$R = \frac{1 + \left(\frac{L_y}{L_z}\right)^2}{4 - 2.521\frac{L_z}{L_y}\left(1 - \frac{1.991}{e^{\pi\frac{L_y}{L_z}} + 1}\right)}\left(1 + \frac{0.00851L_y^2}{L_x^2}\right) - 0.06\left(\frac{L_y}{L_x}\right)^{3/2}\left(\frac{L_y}{Lz} - 1\right)^2 \tag{18}$$

The fundamental resonance frequencies in flexural, longitudinal and torsional mode are identified, and values of Young's and Shear moduli are calculated and listed in the Tables 2 and 3. The measurements are performed several times on a set of several different replicas of composite specimens. The tables contains average values ($\overline{E}$ or $\overline{G}$) calculated from measurements and standard deviation S.

The results from experimental investigations are presented in Fig. 6 in polar plots. The missing data for specimens oriented at 60° are approximated. The Young's modulus is highest in 0° and 90° and smallest in 45° (Fig. 6a). The small difference between values obtained at 0° and 90° is caused probably by technology production feature. The values of shear modulus are highest at 45° and smaller at directions 0° and 90° as can be seen in the Fig. 6b.

4 Conclusion

The results obtained from ultrasonic methods are referred to the properties measured through the thicknesses of the specimens. They do not depend on the orientation of the fiber cloth. The results obtained by static and dynamic impulse excitation methods are in a good agreement for the examined composite beams. The elastic moduli depend on the orientation of fiber cloth warp. The used NDT methods allow identification of anisotropy of material as shown in the Fig. 6. The applied techniques demonstrate convenient, fast and accurate estimation of elastic properties.

References

1. Gross, D.W.: Hauger and All, Engineering Mechanics. Springer. ISBN: 978-3-642-12886-8 (2011)
2. ASTM Standard: Standard Test Method for Flexural Properties of Unreinforced and Reinforced Plastics and Electrical Insulating Materials by Four-Point Bending
3. Autar, K.: Mechanics of Composite Materials. Taylor & Francis Group, Boca Raton, London, New York. ISBN 0-8493-1343-0 (2006)
4. Krautkramer, J.H.: Ultrasonic Testing of Materials. Springer, Berlin (1977)
5. ASTM Standard 1875-00e1: Standard Test Method For Dynamic Young's Modulus, Shear Modulus, and Poisson's Ratio by Sonic Resonance. ASTM International
6. ASTM Standard E 1876–07: Standard Test Method for Dynamic Young's Modulus, Shear Modulus and Poisson's Ratio by Impulse Excitation of Vibration
7. ASTM Standard C 1548–02: Standard Test Method for Dynamic Young's Modulus, Shear Modulus, and Poisson's Ratio of Refractory Materials by Impulse Excitation of Vibration
8. Akishin, P.Y., Barkanov, E.N., Wesolowski, M., Kolosova, E.M.: Static and dynamic techniques for non-destructive elastic material properties characterization. In: Proceedings of XLII International Summer School Conference APM pp. 164–176 (2014)
9. Rao, S.: Vibration of Continuous Systems. Wiley & Sons. (2007)
10. Timoshenko, S.: Vibration Problems in Engineering. D. van Nostrand Company, Inc. 2nd ed, (1937)
11. Radovic, M., Lara-Curzio, E., Riester, L.: Comparison of different experimental techniques for determination of elastic properties of solids. Mat. Sci. Eng. **A368**, 56–70 (2004)
12. Rikards, R., Chate, A., Gailis, G.: Identification of elastic properties of laminates based on experiment design. Int. J. Solids Struct. **38**, 5097–5115 (2001)
13. Spinner, S., Reichard, T.W., Tefft, W.E.: A comparison of experimental and theoretical relations between Young's modulus and the flexural and longitudinal resonance frequencies of uniform bars. J. Res. Nat. Bureau Stand. A Phys. Chem. **64A**, 2, March–April (1960)
14. Spinner, S., Tefft, W.E.: A method for determining mechanical resonance frequencies and for calculating elastic moduli from these frequencies. In: Proceedings ASTM, pp. 1221–1238 (1961)
15. Etcheverry, J., Sánchez, G.: Resonance frequencies of parallelepipeds for determination of elastic moduli: An accurate numerical treatment. J. Sound Vib. **321**(2009), 631–646
16. Etcheverry, J., G. Sánchez, N. Bonadeo, L.I., Raggio: Analysis of the ASTM standards for impulse excitation of vibration and acoustic resonance techniques for rectangular parallelepipeds, VI Congreso Ibero americano de Acústica – FIA 2008 FIA2008-A025
17. Bullough, C. K.: The Determination of Uncertainties in Dynamic Young's Modulus Manual of Codes of Practice for the Determination of Uncertainties in Mechanical Tests on Metallic Materials Code of Practice No. 13 Standards Measurement & Testing Project No. SMT4-CT97-2165
18. Jalili, M.M., Mousavi, S.Y., Pirayeshfar A.S.: Investigating the acoustical properties of carbon fiber-, glass fiber-, and hemp fiber-reinforced polyester composites, polymer composites. **35**(11), 2103–2111 (2014)

Integer Codes for Flash Memories

Hristo Kostadinov and Nikolai L. Manev

Abstract This paper demonstrates the flexibility of integer codes with regard to various type of applications. New constructions of integer codes correcting asymmetric type of errors are proposed in the paper and how to apply the constructed codes to flash memories is discussed.

1 Introduction

Nonvolatile memory is computer memory that maintains stored information without a power supply. For example, the now ancient punch card is a type of nonvolatile memory because, thought it requires power to punch, it does not require power to remain punched. With the rise of portable electronic devices like cell phones, mp3 players, digital cameras, and PDAs, nonvolatile memory is increasingly important. Flash memory is currently the dominant nonvolatile memory because it is cheap and, unlike punch cards and other more recent kinds of nonvolatile memory, can be electrically programmed and erased with relative ease.

A chip of flash memory contains an array of tens of thousands of cells, and we assume that each chip stores a bit string. Each cell on a chip of flash memory can be thought of as a container of electrons. In binary flash each cell has two states: if there are electrons in the container then the cell is in the state 1, and if there are no electrons in the container, the cell is in state 0. Until recently, binary flash was the only kind of flash available, but now a new kind of flash memory has been developed, multilevel flash, that many see as the future of flash memory. In a multilevel cell, it is possible to distinguish between several different ranges of charge, allowing for more than two states.

H. Kostadinov (✉)
IMI-BAS, Acad. G. Bonchev St., Bl.8, 1113 Sofia, Bulgaria
e-mail: hristo@math.bas.bg

N.L. Manev
USEA "Lyuben Karavelov" and Institute of Mathematics and Informatics,
IMI-BAS, Acad. G. Bonchev St., Bl.8, 1113 Sofia, Bulgaria
e-mail: nlmanev@math.bas.bg

K. Georgiev et al. (eds.), *Advanced Computing in Industrial Mathematics*, Studies in Computational Intelligence 728,
https://doi.org/10.1007/978-3-319-65530-7_9

To transition between the states, it is necessary to add and remove electrons to and from the container. While it is easy to add electrons (i.e. to increase the state of the cell), it is impossible to remove electrons (i.e. to decrease the state of the cell) without first emptying the electrons from all the containers in a large selection of the chip. This process, called reset operation, is slow and, after many repetitions, wears out the chip. In multilevel flash, there are two types of mistakes that can occur when programming a cell: errors in which too many electrons are added ("overshoots") and errors in which too few electrons are added ("undershoots"). Because of the difficulty of removing electrons, overshoots are much bigger problem than undershoots. To avoid overshoots, the level of a cell is increased over multiple iterations by carefully adding small number of electrons at a time.

Flash devices exhibit a multitude of complex error types and behaviors, but common to all flavors of flash storage is the inherent asymmetry between cell programming (charge replacement) and cell erasing (charge removal). This asymmetry causes significant error sources to change cell levels in one dominant direction. Moreover, many reported common flash error mechanisms induce errors whose magnitudes (the number of error changes) are small, and independent of the alphabet size, which may be significantly larger than the typical error magnitude. In addition to the (uncontrolled) errors that challenge flash memory design and operation, codes for asymmetric limited-magnitude errors can be used to speed-up the memory access by allowing less-precise programming schemes that introduce errors in a controlled way. While not a panacea for all flash issues, the potential error migration and performance boost by asymmetric limited-magnitude codes, justify their addition, alongside other coding innovations, to the menu of flash coding solutions.

The most well-studied model for error-correcting codes is the model for symmetric errors. According to this model, a symbol, taken from the code alphabet, is changed to another symbol from the same alphabet, and all such are equally likely. The popularity of this model stems from both its applicability to a broad set of applications, and from the powerful construction techniques that were found to address it. In addition to the symmetric model, many other models, variations and generalizations were studied, each motivated by a behavior of practical systems or applications.

The asymmetric limited-magnitude error correcting codes can be used to speed up the writing process to flash devices (memory write is referred to as programming in the flash literature). This is done by relaxing the programming accuracy requirements, and using the codes to correct the resulting programming errors. Since the flash programming mechanism is inherently probabilistic, the introduction of "intentional" programming errors in a controlled way can significantly reduce the average programming time and improve the write performance. Such an outcome would be highly desirable given the inferiority of flash devices in write performance compared to their read performance, and to the sequential write performance of the hard-disk devices.

Asymmetric limited-magnitude error-correcting codes were proposed in [1]. The codes, proposed in that paper, were for the special case of correcting all asymmetric limited-magnitude errors within the codeword. These codes turn out to be a special case of the general construction method provided by Cassuto et al. [2].

In 2011, Klove and Bose [3] proposed systematic codes that correct single limited-magnitude systematic asymmetric errors and achieve higher rate than the ones given in [2]. They also showed how their code construction can be slightly modified to gives codes correcting symmetric errors of limited magnitude. Later Klove et al. [4] extended their result and gave a necessary and sufficient condition for existing a code over $GF(p)$ correcting a single asymmetric error.

As it has been already mentioned, asymmetric errors in flash memories are very common. However, there are cases in which the possible error type includes both a symmetric and an asymmetric error. For example, let us have a flash memory with n voltage levels and have to increase the voltage level of a cell with current level $t-1$ by one (which is an usual situation when programming a flash memory). In such a case the most common error observed is overcharging the cell (increasing the level with at least 2, or to charge it less than is needed, i.e. after charging the cell stays at level $t-1$. Hence, that kind of error is a combination of the symmetric error (± 1) and the asymmetric error $(2, 3, \ldots, n)$.

The next of the paper is organized as follows. The necessary notations and definitions which are used in this paper are given in Sect. 2.1. In Sect. 2.2 we briefly discuss some existing results. New construction of integer codes for flash memory are presented in Sect. 3. Conclusion remarks and some open problems are discussed in Sect. 4.

2 Preliminaries

2.1 Integer Codes

Asymmetric error correcting codes were introduced by Varshamov and Tenegolz [5] in the middle of 60s. In that work they also gave the definition of integer code. For many years these codes have been almost forgotten. The appearance of multilevel flash memories renewed the interest in codes correcting asymmetric errors.

Integer codes are codes defined over finite rings of integers. Han Vinck and Morita [6] investigated integer codes with a view to magnetic recording and frame synchronization. A class of integer codes correcting specific types of errors and their application to coded modulation has been proposed by Kostadinov et al. [7]. Because of their flexibility integer codes are very suitable for application in multilevel flash memory.

Definition 1 Let $\mathbb{Z}_A$ be the ring of integers modulo A. An ***integer code*** of length n with parity-check matrix $\mathbf{H} \in \mathbb{Z}_A^{m \times n}$, is referred to be a subset of $\mathbb{Z}_A^n$, defined by

$$\mathscr{C}(\mathbf{H}, \mathbf{d}) = \{\mathbf{c} \in \mathbb{Z}_A^n \mid \mathbf{c}\mathbf{H}^T = \mathbf{d} \mod A\}$$

where $\mathbf{d} \in \mathbb{Z}_A^m$.

If $d = 0$ the code is a linear $[n, n-m]$ code over $\mathbb{Z}_A$. Without loss of generality, we can assume $d = 0$ in this paper. We write $\mathscr{C}(\mathbf{H})$, or only $\mathscr{C}$ if there is no possibility for ambiguity.

In this paper we consider codes with $m = 1$ (one check symbol only). Then $\mathbf{H} = (h_1, h_2, \ldots, h_n)$, $0 \neq h_i \in \mathbb{Z}_A$ and

$$\mathscr{C}(\mathbf{H}) = \{\mathbf{c} \in \mathbb{Z}_A \mid \sum_{i=1}^{n} c_i h_i = 0 \mod A\}$$

Integer codes are designed to correct specific type of errors instead of correcting number of errors in a codeword as it is the case with conventional codes. Thus, we need the following definition.

Definition 2 Let l_j and e_i be positive integers, $j = 1, \ldots, m,\ i = 1, \ldots, s$. The code $\mathscr{C}(\mathbf{H}, d)$ is said to be a ***single*** $(l_1, l_2, \ldots, l_m, \pm e_1, \pm e_2, \ldots, \pm e_s)$***-error correctable*** if it can correct any single error with value l_j or $\pm e_i$.

Obviously, $\mathscr{C}(\mathbf{H}, d)$ is a single $(l_1, l_2, \ldots, l_m, \pm e_1, \pm e_2, \ldots, \pm e_s)$-error correct-able code if and only if the subsets $\{h_j\, l_1, h_j\, l_2, \ldots, h_j\, l_m, \pm h_j\, e_1, \pm h_j\, e_2, \ldots, \pm h_j\, e_s\} \subset \mathbb{Z}_A$, are pairwise disjoint and of the same cardinality $2s + l$, for any $j = 1, 2, \ldots, n$. Thus, we have

$$A \geq (2s + l)n + 1.$$

Definition 3 A single $(l_1, l_2, \ldots, l_m, \pm e_1, \pm e_2, \ldots, \pm e_s)$-error correctable code $\mathscr{C}(\mathbf{H}, d)$ of block length n is called ***perfect***, when $A = (2s + l)n + 1$.

In most of the cases perfect integer codes do not exist. We shall say that a single $(l_1, l_2, \ldots, l_m, \pm e_1, \pm e_2, \ldots, \pm e_s)$-error correctable integer code $\mathscr{C}(\mathbf{H}, d)$ of block length n over $\mathbb{Z}_A$ is ***optimal*** if A is the minimum value for which the code $\mathscr{C}(\mathbf{H}, d)$ exists.

Remark One side effect, however, is that part of the power of the integer codes is used to correct wrap-around errors (i.e. errors modulo A), which does not appear in the flash memories. More precisely, we assume that a codeword c may be changed into $c+e \pmod A$. If $c+e < 0$ or $c+e \geq A$, these are ***wraparound errors***. However, such errors usually constitute a minor part of the correctable errors. We can estimate this effect by a heuristic argument and show that when A is large compared to the maximum value of the set $\{l_i, \pm e_j\}$, where $i = 1, \ldots, m$ and $j = 1, \ldots, s$, the main power of the code can be used to correct errors in flash memories.

2.2 Several Proposed q-ary Codes

In [2] Cassuto et al. describe a general method of constructing t-asymmetric λ-limited-magnitude error correcting codes from codes correcting symmetric errors.

Recently Klove et al. in [3] and [4] have done thorough study of t-asymmetric λ-limited-magnitude error correcting codes over $\mathbb{Z}_A$. Their study is based on the fact that the discussed coding problems can be reformulated and solve as problems in number theory.

Definition 4 An error vector $\mathbf{e} = (e_1, e_2, \ldots, e_n)$ is called a ***t-asymmetric λ-limited-magnitude error*** if $\mathrm{wt}(\mathbf{e}) = |\{i : e_i \neq 0\}| \leq t$ and $0 \leq e_i \leq \lambda$, for all $i = 1, 2, \ldots, n$. A code $\mathscr{C}$ is called a ***t-asymmetric λ-limited-magnitude error correcting code*** if it can correct all t-asymmetric λ-limited-magnitude errors.

Let $E(\lambda, n, t)$ denote the set of all possible t asymmetric λ-limited-magnitude error vectors of length n over A.

In the cited papers the notation $B_t[\lambda](A)$ is used, or just $B_t[\lambda]$ when A is known from the context. Namely, $B_t[\lambda](A)$ is defined as a set $B_t[\lambda](A) = \{b_1, b_2, \ldots, b_n\}$ such that the set

$$\mathbf{e}B_t[\lambda](A) = \left\{ e_1 b_1 + e_2 b_2 + \cdots + e_n b_n \mid \mathbf{e} \in E(\lambda, n, t) \right\}$$

consists of distinct elements of $\mathbb{Z}_A$, i.e., modulo A. In these papers classes of codes correcting $t = n$ and $t = n - 1$ asymmetric λ-limited-magnitude errors are proposed. But the most attention was paid to the case $t = 1$, i.e., the set $B_1[\lambda](A)$. The Hamming bound for such codes gives $A \geq 1 + \lambda n$.

Define $M_\lambda(A)$ to be the maximal size of a $B_1[\lambda](A)$ set. In [3] it has been shown that for odd values of A we have

$$M_\lambda(A) = \frac{A-1}{2} - \frac{\omega_A}{2}$$

where ω_A is the number of the cyclotomic cosets of odd size. In [4] $M_2(A)$ and bounds for $M_3(A)$ and $M_4(A)$ are determined.

In [4] a perfect $B_1[\lambda](p)$ sets for a class of primes p is described. Also some results about $B_1[\lambda](A)$, $\lambda = 3, 4$, are obtained. Unfortunately theoretical results gives good codes for very large values of A. Optimal for codes over reasonable large alphabets are found by computer search in the case $t = 2$ and $t = n - 2$ for small n.

3 New Constructions of Integer Codes Correcting Single Type of Errors

In this Section we propose two constructions of integer codes correcting single errors. The next theorem gives the exact form of the check matrix of an integer code correcting a single asymmetric 2-limited-magnitude error.

Theorem 1 *A 1-asymmetric 2-limited-magnitude error correctable code $\mathscr{C}$ of length n over Z_A has the following parity-check matrix* $\mathbf{H}$

- $\mathbf{H} = (1, 3, 5, \ldots, n-1, n+3, n+5, \ldots, 2n+1)$, *where* $A = 2n+2$ *and* n *is even*
- $\mathbf{H} = (1, 3, 5, \ldots, n-2, n+4, n+6, \ldots, 2n+3)$, *where* $A = 2n+4$ *and* n *is odd*

Remark In the case when n is even the code is "almost" perfect—the exceeding is 1.

Proof Here we are going to prove the case when n is even and $A = 2n+2$. The proof when n is odd is analogous.

To show that a code C with parity-check matrix

$$\mathbf{H} = (1, 3, 5, \ldots, n-1, n+3, n+5, \ldots, 2n+1)$$

is 1-asymmetric 2-limited-magnitude error correctable it is enough to prove that all elements of $\mathbf{H}_1 = 2\mathbf{H} \pmod{2n+2}$ are distinct and $\mathbf{H} \cap \mathbf{H}_1 = \emptyset$. We have

$$2H = (2, 6, 10, \ldots, 2n-10, 2n-6, 2n-2, 2n+6, 2n+10, \ldots, 4n-2, 4n+2)$$

and

$$H_1 = (2, 6, 10, \ldots, 2n-6, 2n-2, 4, 8, \ldots, 2n-4, 2n).$$

It is not so difficult one to see that all the elements in $\mathbf{H}_1$ are distinct. Moreover, the elements of $\mathbf{H}_1$ are even, while the elements of $\mathbf{H}$ are odd. So we have $\mathbf{H} \cap \mathbf{H}_1 = \emptyset$. With that the proof is completed.

Let P_o be the set of odd primes p such that $\mathrm{ord}_p(2)$ is odd. And let $A = 2n+2$ and $p|(A-1)$ where $p \in P_o$. According to Theorem 2 [4], it does not exist a 1-asymmetric 2-limited-magnitude error correctable code of length n over Z_{A-1}. So, we can construct a 1-asymmetric 2-limited-magnitude error correctable code of length n over Z_A using Theorem 1, which is quasi-perfect. In such a way, we improve the result given in [4] in case of the length of the code n such that $p|(2n+1), p \in P_o$.

Now we shall investigate how to construct an integer code $\mathscr{C}(\mathbf{H})$ capable to correct a single error of type $(\pm 1, 2)$. Because the code will be single error correctable, its check matrix $\mathbf{H}$ has to consist of a single row.

First, let us consider the set of integers

$$B = B(m) = \{4^k l < m \mid k, l, m \in \mathbb{N},\ m \geq 6 \text{ is even, and } l \text{ is odd}\}.$$

Let us divide the set B into two subsets—B_0 and B_1, where

$$B_0 = \{a \in B \mid \exists\, b \in B : \ 2a + b \equiv 0 \pmod{2m}\} \quad \text{and} \quad B_1 = B \setminus B_0. \quad (1)$$

Remark Since $0 < a, b < m$ then $2a + b \equiv 0 \pmod{2m}$ is equivalent to $2a + b = 2m$. Hence $2a = 2m - b \leq 2m - 4$, i.e., $a \leq m - 2$. On the other hand $2m - 2a = b < m$ gives $a > m/2$. Therefore,

$$\frac{m}{2} < a \leq m - 2.$$

But not all integers in the above interval belongs to B_0. It is not difficult to prove that for $m = 2^k$ we have $B_0 = \emptyset$.

Example 1 Let $m = 82$. Following the definition of B, B_0 and B_1 we obtain

$$B = \{4, 12, 16, 20, 28, 36, 44, 48, 52, 60, 64, 68, 76, 80\}$$

$$B_0 = \{44, 48, 52, 60, 64, 68, 76, 80\}$$

and

$$B_1 = \{4, 12, 16, 20, 28, 36\}.$$

We have the following construction for a single $(\pm 1, 2)$ error correctable integer code.

Theorem 2 *Let* $m \geq 6$ *is a given integer and* m *is even. Let us consider the sets* $B(m), B_0$ *and* B_1. *The integer code* $\mathscr{C}(\mathbf{H})$ *over* Z_{2m} *with the check matrix*

$$\mathbf{H} = (1, 3, 5, 7, \ldots, m-1 \mid B_1)$$

is a single $(\pm 1, 2)$ *error-correctable.*

Proof The integer code $\mathscr{C}(\mathbf{H})$ is a single $(\pm 1, 2)$ error-correctable if all its syndrome values are different. Hence, to prove the theorem will be enough to show that

$$\mathbf{H} \cap (-\mathbf{H}) \cap (2\mathbf{H}) = \emptyset, \tag{2}$$

where all the operation are taken into Z_{2m}.

For convenience, let us divide $\mathbf{H}$ into 2 subsets $A_1 = (1, 3, 5, 7, \ldots, m-1)$ and B_1. So, the Eq. (2) is equivalent to

$$A_1 \cap (-A_1) \cap (2A_1) \cap B_1 \cap (-B_1) \cap (2B_1) = \emptyset. \tag{3}$$

One can easily see that $-A = (m+1, m+3, m+5, \cdots, 2m-1)$, and $A_1 \cap (-A_1) \cap (2A_1) = \emptyset$. Moreover,

$$A_1 \cup (-A_1) = \{2n + 1 | n = 0, 1, 2, 3 \ldots, m-1\} \tag{4}$$

and

$$2A_1 = \{4n + 2 | n = 0, 1, 2, 3 \ldots, m/2 - 1\} \tag{5}$$

On the other side, $2m$ is divisible by 4. Hence, all the elements of the sets $B_1, -B_1 = \{2m - b | b \in B_1\}$ and $2B_1$ are divisible by 4. So, using (4) and (5) we have

$$(A_1 \cup (-A_1) \cup (2A_1)) \cap (B_1 \cup (-B_1) \cup (2B_1)) = \emptyset. \qquad (6)$$

The only thing that we have to show is that

$$B_1 \cup (-B_1) \cup (2B_1) = \emptyset. \qquad (7)$$

It is obvious that $B_1 \cup (2B_1) = \emptyset$, because all the elements of B_1 are not divisible by 8, while all the elements of $2B_1$ are divisible by 8. We have that $B_1 \cup (-B_1) = \emptyset$, since $2m - b_i > b_j$, where $b_i, b_j \in B_1$.

To prove that $(-B_1) \cup (2B_1) = \emptyset$ we should show that $2a + b \neq 0 \pmod{2m}$, where $a, b \in B_1$. But that follows from (1) and the definition of the set B_1. Hence, using (6) and (7) we complete the proof of the theorem.

Example 2 Let $m = 64$. For the sets B, B_0 and B_1 we have

$$B_m = \{4, 12, 16, 20, 28, 36, 44, 48, 52, 60\}, \qquad B_0 = \emptyset,$$

$$B_1 = \{4, 12, 16, 20, 28, 36, 44, 48, 52, 60\}.$$

So, the integer code $\mathscr{C}(\mathbf{H})$ over Z_{128} with the check matrix

$$\mathbf{H} = (1, 3, 5, 7, \dots, 63, 4, 12, 16, 20, 28, 36, 44, 48, 52, 60)$$

is a single $(\pm 1, 2)$ error-correctable. The length of the code is 42 and it is optimal. We can say that the code is "almost" perfect, because the exceeding is only 1.

Let $a \in B_0$, $b \in B_1$ and $2a + b \equiv 0 \pmod{2m}$. It is easy to see that if we change the elements b with a in $\mathbf{H}$ the theorem still holds.

4 Conclusion

In this work we have presented two new constructions of single error correctable integer codes designed for an application in a flash memory. Moreover, we gave the exact form of the check matrix for those codes. For some parameters, the obtained codes are optimal. The decoding complexity is linear, regarding to the code length, and can be used a look-up table to decode them. All these advantages of integer codes makes them very suitable for their usage in the practice. One can see that we only consider the case of single error and small magnitude. Actually, it is very difficult to obtain theoretical results for multiple errors and higher magnitude. On that we will focus for our future research [8].

Acknowledgements This work was partially supported by the National Science Fund of Bulgaria under Grant DFNI-I02/8.

References

1. Ahlswede, R., Aydinian, H., Khachatrian, L.: Undirectional error control codes and related combinatorial problems. In: Proceedings of the Eighth International Workshop on Algebraic and Combinatorial Coding Theory (ACCT0'2002), pp. 6–9 (2002)
2. Cassuto, Y., Schwartz, M., Bohossian, V., Burck, J.: Codes for asymmetric limited-magnitude errors with application to multi-level flash memories. IEEE Trans. Inf. Theory **56**(4):1582–1595 (2010)
3. Klove, T., Bose, B.: Sistematic, single limited magnitude error correcting codes for flash memories. IEEE Trans. Inf. Theory **57**(7):4477–4487 (2011)
4. Klove, T., Lou, J., Naydenova, I., Yari, S.: Some codes correcting asymmetric errors of limited magnitude. IEEE Trans. Inf. Theory **57**(11):7459–7472 (2011)
5. Varshamov, R., Tenegolz, G.: One asymmetrical error-correctable codes. Automatika i Telematika **26**(2):288–292 (1965)
6. Han Vinck, A.J., Morita, H.: Codes over the ring of integers modulo *m*. IEICE Trans. Fundam. **E81-A**(10):2013–2018 (1998)
7. Kostadinov, H., Morita, H., Manev, N.: Integer codes correcting single errors of specific types $(\pm e_1, \pm e_2, \ldots, \pm e_s)$. IEICE Trans. Fundam. **E86-A**:1843–1849 (2003)
8. Kostadinov, H., Manev, N.: On codes for flash memories. In: International Workshop on Algebraic and Combinatorical Coding Theory (ACCT), pp. 197–202. Pomorie, Bulgaria (2012)

On the Strong Asymptotics of Rows of the Padé Table

Ralitza K. Kovacheva

Abstract In the present paper, results on the strong asymptotics of row sequences $\{\pi_{n,m}\}$, $n \to \infty$, m-fixed of classical Padé approximants are provided.

Keywords Padé approximation · Strong convergence

MSC: 41A21 · 41A25 · 30E10

1 Introduction

Let $f(z) := \sum_{j=0}^{\infty} f_j z^j$ be a function holomorphic (single valued or a single valued branch of an analytic function) at the origin. Suppose that the radius of holomorphy R_0 is finite and set

$$(f(z))_{(n)} := S_n(z) = \sum_{j=0}^{n} f_j z^j.$$

As it is known (see, for instance [10] and [21]),

$$\limsup_{n\to\infty} |S_n(z)|^{1/n} = \frac{|z|}{R_0} \text{ for } |z| > R_0$$

The formula above refers to the weak asymptotics of the sequence $\{S_n\}_{n\to\infty}$.

Furthermore, a strong asymptotics takes place. More exactly, there exists an infinite sequence Λ of positive integers such that

$$\frac{S_n(z)}{f_n z^n} \to \chi(z),\ n \to \infty,\ n \in \Lambda$$

R.K. Kovacheva (✉)
Institute of Mathematics and Informatics, Bulgarian Academy of Sciences,
Acad. Bonchev str. 8, 1113 Sofia, Bulgaria
e-mail: rkovach@math.bas.bg

K. Georgiev et al. (eds.), *Advanced Computing in Industrial Mathematics*, Studies in Computational Intelligence 728,
https://doi.org/10.1007/978-3-319-65530-7_10

uniformly in the *max*-norm on compact subsets of $D^c_{R_0}$, where $\chi(z)$ is a function analytic in $D^c_{R_0}$ and $\chi \not\equiv 0$. ([12]); $D_{R_0} := \{z, |z| < R_0\}$.

In what follows, we will write "uniformly inside" $D^c_{R_0}$.

Given a pair of integers (n, m), let $\pi_{n,m}$ be the classical Padé approximant of f of order (n, m). Recall that $\pi_{n,m} = p/q$, where p, q are polynomials of degree $\leq n, m$ respectively and such that

$$(fq - p)(z) = 0(z^{n+m+1}).$$

As it is well known (see [15]), the Padé approximant $\pi_{n,m}$ always exists and is uniquely determined by the condition above.

For the sake of accuracy, we recall that the sequence $\{\pi_{n,m}\}, n \to \infty, m$-fixed, is called the "$m$th row of the Padé table, associated with the power series f."

In the present paper, we pose the question about the strong asymptotics of rows of classical Padé approximants.

2 Statement of the Results

In the forthcoming consideration, the integer m is fixed.

Set

$$\pi_{n,m} := \pi_n = \frac{P_n}{Q_n},$$

where $(P_n, Q_n) = 1$ and Q_n is monic.

Let

$$f(z) - \pi_n(z) = 0(z^{n+m+1-\tau_n}),\ \tau_n \geq 0.$$

Apparently, $\deg P_n \leq n - \tau_n$, $\deg Q_n \leq m - \tau_n$. Denote by A_n the leading coefficient of the polynomial P_n, that is:

$$P_n(z) = A_n z^{\deg P_n} + \cdots.$$

Before continuing, we introduce the notation $\mathcal{M}_m(\cdot)$. Given a point set B in the complex plane $\mathbf{C}$, $\mathcal{M}_m(B)$ will stand for the class of functions, meromorphic and having no more than m poles in B (poles are counted with regard to their multiplicities). Further, we adopt the notation $\mathcal{A}(B)$ for functions, holomorphic in B.

Set $R_m(f) := R_m$ for the radius of m-meromorphy of the function f, that is:

$$R_m := \sup_R \{f \in \mathcal{M}_m(D_R)\}.$$

Apparently, $R_m \geq R_0$. Furthermore, $R_m > 0$ ensures that $R_0 > 0$ ([7]).

Our main result is

Theorem 1 *Given a power series* $f(z) := \sum_{j=0}^{\infty} f_j z^j$ *and* m *a fixed integer, suppose that* $0 < R_m < \infty$. *Assume that* f *is analytic on the circle* C_{R_m}, *except for a single pole at the point* a, $|a| = R_m$.

Then there is an infinite sequence Λ *and a function* $\chi \in \mathscr{A}(D^c_{R_m})$, $\chi \not\equiv 0$ *such that*

$$\frac{P_n(z)}{A_n z^n} \to \chi(z),\ n \in \Lambda$$

uniformly inside $D^c_{R_m}$.

A natural consequence of Theorem 1 is a result of Jentzsch-Szegö type about the asymptotic distribution of the zeros of the polynomial sequence $\{P_n\}$ as $n \in \Lambda$. Before formulating it, we introduce the term of the "counting measure" μ_P of a polynomial P, that is:

$$\mu_P(K) := \frac{\nu_P(K)}{\deg P},$$

where $\nu_P(K)$ stands for the number of the zeros of P on the set K. Further, a sequence of measures $\{\mu_n\}$ supported by a set S converges weakly to the measure μ iff

$$\int g\, d\mu_n \to \int g d\mu$$

for every continuous function in the complex plane (for details, the reader is referred to [14]).

Corollary 2 *Under the conditions of Theorem 1,*

$$\mu_{P_n} \longrightarrow^* \mu_{R_m} \ as\ n \in \Lambda.$$

Here "$\longrightarrow^*$" stands for the weak convergence of the counting measures μ_{P_n} as $n \in \Lambda$ and μ_{R_m} is the equilibrium measure for the disk $\overline{D_{R_m}}$; in the case being considered, $\mu_{R_m} = \frac{1}{2\pi R_m} ds_{R_m}$, look at [17]).

3 Proofs

The proofs will be preceded by auxiliary lemmas.

Lemma 1 (Kakehashi's Regularization Lemma), [11] *Let* $\{c_n\}$ *be an infinite sequence of complex numbers such that*

$$\limsup_{n\to\infty} |c_n|^{1/n} = c \in (0, \infty).$$

Then there exists a monotone sequence $\{\lambda_n\}$ *such that*

(1) $\lim_{n\to\infty} \lambda_{n+1}/\lambda_n \to 1$ *as* $n \to \infty$;
(2) *setting* $c_n^* := \lambda_n c^n$, *we have*

$$\begin{cases} c_n^* \geq |c_n| \text{ for every } n \\ |c_n| = c_n^* \text{ for a subsequence } \Lambda, \end{cases}$$

Of fundamental importance for the coming considerations is the classical theorem of Montessus de Ballore, which we present as

Lemma 2 (Montessus de Ballore), [6] *Let* f *be a power series holomorphic at the zero and* m *be fixed. Assume that* $R_m < \infty$ *and that* f *has exactly* m *poles in* D_{R_m} *say, at the points* $a_1, ..., a_m$. *Then the sequence* $\{\pi_{n,m}\}$ *converges uniformly to* f *inside* $D_{R_m} - -\{a_j\}_{j=1}^m$ *and each pole of* f *attracts, as* $n \to \infty$, *as many free poles of* Q_n *as its multiplicity.*

The classical Montessus de Ballores theorem provoked a new approach to the research on meromorphic continuation of functions and characterization of their singularities (see [4, 5, 7, 9, 13, 18–20]), as well as to investigations connected with the rate of approximation with rational functions and asymptotic distribution of the zeros of the approximating sequences (see [1, 2, 8]).

Lemma 3 (Blatt-Saff-Simkani), [3] *Let* E *be a compact set in the complex plane with positive Green's capacity* $capE$ *and let* μ_E *be the equilibrium measure for* E. *Let* Λ *be a sequence of positive integers and* $\{p_n\}$ *be monic polynomials of respective degrees precisely* n. *Assume that*

(a) $\limsup_{n\in\Lambda} ||p_n||_E^{1/n} \leq capE$ *and*
(b) $\lim_{n\in\Lambda} \mu_n(A) = 0$ *for every closed set* A *contained in the union of the bounded components of the complement of the set* E *in the extended complex plane.*

Then $\mu_n \longrightarrow^* \mu_E$ *as* $n \in \Lambda$.

3.1 Proof of Theorem 1

Proof Under the conditions of the theorem, the function f has exactly m poles in D_{R_m}, say $\alpha_1, ..., \alpha_m, 0 \leq |\alpha_k| \leq |\alpha_{k+1}| < R_m, k = 1, ..., m-1$ and a single pole at the point $a \in C_{R_m}$. In view of Lemma 2,

$$\begin{aligned} &Q_n(z) \to Q(z) := (z-\alpha_1)...(z-\alpha_m),\ n \to \infty, \deg Q_n = m, n \geq n_0. \\ &\text{and} \\ &(f - \pi_n)(z) = O(z^{n+m+1}). \end{aligned} \tag{1}$$

Set $F := fQ$. Clearly, $F \in \mathscr{A}(D_{R_m})$ and has a single pole at $a \in C_{R_m} := \partial D_{R_m}$.

By (1),

$$(QP_n) = (FQ_n)_{(n+m)}. \tag{2}$$

Set

$$Q_n(z) := a_{n,m}z^m + a_{n,m-1}z^{m-1} + \cdots + a_{n,0},\ a_{n,m} = 1.$$

Under the conditions of the theorem,

$$|a_{n,k}| \le C_1,\ k = 0, \ldots, m,\ n \ge n_1 \ge n_0. \tag{3}$$

In what follows, we denote by $C_j, j = 1, 2, \ldots$ positive constants not depending on n.

We have

$$(FQ_n)_{(n+m)}(z) = \sum_{l=0}^{m} z^l \Big(\sum_{k=0}^{l} F_{l-k}a_{n,k}\Big) + \sum_{l=m+1}^{m+n} z^l \Big(\sum_{k=0}^{m} F_{l-k}a_{n,k}\Big). \tag{4}$$

Set

$$(FQ_n)_{(n+m)}(z) := \sum_{l=0}^{n+m} B_{n,l}z^l.$$

Let $\{F^*_n\}$ be Kakehashi's regularization of the sequence $\{F_n\}$, $F^*_n = \lambda_n/R^n_m$, and the infinite sequence Λ be such that $|F_{n+m}| = F^*_{n+m}$, $n \in \Lambda$. In the sequel, we shall be dealing with $n \in \Lambda$.

Notice, that by Lemma 2,

$$\lim_{n\in\Lambda} |F_{n+m}|^{1/n} = 1/R_m. \tag{5}$$

Combining (2) and (3), we get

$$\frac{|B_{n,l}z^l|}{F^*_{n+m}|z^{n+m}|} \le$$

$$C_1 \sum_{k=0}^{l} \Big|\frac{F^*_{l-k}}{F_{n+m}}\Big| |z|^{l-m-n} = C_1 \sum_{k=0}^{\min(l,m+1)} \frac{\lambda_{l-k}}{\lambda_{n+m}} R_m^{n+m-l+k} |z|^{l-m-n}. \tag{6}$$

Fix now an arbitrary positive number ε. The monotony of the sequence $\{\lambda_j\}$ ensures that

$$\frac{\lambda_j}{\lambda_{j+1}} \le 1 + \varepsilon$$

for every j great enough, say $j \ge j_0$. Without losing the generality, we suppose that $j_0 > m$.

The last inequality yields, together with (6)

$$\left|\frac{B_{n,l}z^l}{F_{n+m}z^{n+m}}\right| \leq C_2 \frac{(1+\varepsilon)R_m)^{n+m-l}}{|z|^{n+m-l}} \tag{7}$$

which in turn leads to

$$\left|\frac{(FQ_n)_{(n+m)}(z)}{F_{n+m}z^{n+m}}\right| \leq C_2 \sum_{l=0}^{n+m} \left(\frac{(1+\varepsilon)R_m}{|z|}\right)^{n+m-l}, \tag{8}$$

where $C_2 = C_2(\varepsilon)$.

The sequence of the rational functions $\{(FQ)_{(n+m)}(z)/F_{n+m}z^{n+m}\}_{n\in\Lambda}$ is uniformly bounded inside $D^c_{R_m(1+\varepsilon)}$. By the compactness principle, $(FQ)_{(n+m)}(z)/F_{n+m}z^{n+m}$ converges, as $n \in \Lambda$, uniformly inside $D^c_{R_m(1+\varepsilon)}$ to an analytic function. Taking into account (2), as well as the arbitrariness of ε, we obtain

$$\frac{Q(z)P_n(z)}{F_{n+m}z^{n+m}} \to \chi(z) \text{ as } n \to \infty, n \in \Lambda$$

uniformly inside $D^c_{R_m}$, with $\chi \in \mathscr{A}(D^c_{R_m})$. From here, we easily get (see Lemma 2) that

$$\left\{\frac{P_n}{F_{n+m}z^n}\right\} \to \chi \text{ as } n \in \Lambda$$

uniformly inside $D^c_{R_m}$.

In what follows, we show that $\chi \not\equiv 0$. Indeed, by construction,

$$\frac{A_n}{F_{n+m}} = \left(a_{n,m}\frac{F_n}{F_{n+m}} + a_{n,m-1}\frac{F_{n+1}}{F_{n+m}} + \ldots + a_{n,m}\right).$$

Now, we take notice that the function F has a single pole at $a \in C_{R_m}$, hence (see Lemma 2)

$$\frac{F_j}{F_{j+1}} \to a, \text{ as } j \to \infty.$$

Consequently,

$$\lim_{n\to\infty, n\in\Lambda} \frac{A_n}{F_{n+m}} = Q(a) \neq 0. \tag{9}$$

On the other hand,

$$\chi(\infty) = \lim_{n\to\infty, n\in\Lambda} \frac{A_n}{F_{n+m}}.$$

Therefore,

$$\chi(\infty) = Q(a) \neq 0,$$

which ensures that

$$\chi(\infty) \not\equiv 0.$$

On this, Theorem 1 is proved. □

3.2 *Proof of Corollary 2*

Proof The proof is based on Lemma 3. We first remind that $\mathrm{cap}(\overline{D}_R) = R$ (see [17]).
Set μ_n for the counting measures associated with the polynomials P_n, $n \in \Lambda$.
Combining (5) and (9), we obtain

$$\limsup_{n\to\infty, n\in\Lambda} |A_n|^{1/n} = 1/R_m \tag{10}$$

Then the application of Lemma 2 to the polynomials $\{P_n\}$ yields

$$||P_n||_{\overline{D}_{R_m}}^{1/n} \leq 1$$

which, by (10) leads to

$$\limsup_{n\in\Lambda} ||\frac{P_n}{A_n}||_{\overline{D}_{R_m}}^{1/n} \leq R_m.$$

On the other hand, by Hurwitz theorem

$$\lim_{n\to\infty} \mu_n(K) = 0$$

on each compact subset K of D_{R_m}.

Herewith conditions (a) and (b) of Lemma 3 are fulfilled and the statement of the corollary is proven. □

4 Conclusions

Given a power series $f(z) := \sum_{n=0}^{\infty} f_n z^n$, holomorphic at the origin, suppose that f admits a continuation as a meromorphic function with exacty $m+1$ poles at $\alpha_1, \ldots, \alpha_m, \alpha_{m+1}$, where $0 < |\alpha_1| \leq \cdots \leq |\alpha_m| < |\alpha_{m+1}|$ (poles are counted with regard to multiplicities). Under this assumption, the Padé approximants $\{\pi_{n,m}\}$, m-fixed, behave outside $\{z, |z| > |\alpha_{m+1}|\}$ asymptotically through an infinite sequence of natural numbers like the Taylor sums of the function fQ, where $Q(z) := \prod_{j=1}^{m} (z - \alpha_j)$.

References

1. Blatt, H.P., Kovacheva, R.K.: Growth behavior and zero distribution of rational functions. Constr. Approx. **34**(3), 393–420 (2011)
2. Blatt, H.P., Grothmann, R., Kovacheva, R.K.: Regions of meromorphy and value distribution of geometrically converging rational approximants. J. Math. Anal. Appl. **382**, 66–76 (2011)
3. Blatt, H.P., Saff, E.B., Simkani, M.: Jentzsch-Szegö type theorems for the zeros of best approximants. J. Lond. Math. Soc. **38**, 307–316 (1988)
4. Buslaev, V.I.: Relations for the coefficients, and singular points of a function. Mat. Sb. **131**(173), 357–384 (1986) [English transl. in Math. USSR Sb. 59 (1988)]
5. Cacoq, J., de la Calle Ysern, B., López Lagomasino, G.: Direct and inverse results on row sequences of Hermite-Paé approximants. Constr. Approx. **38**, 133–160 (2013)
6. de Montessus de Ballore, R.: Sur le fractions continues algebriques. Bull. Soc. Math. France **30**, 28–36 (1902)
7. Gonchar, A.A.: Poles of the rows of the Padé table and meromorphic continuation of functions. Mat. Sb. **115**(157), 590–615 (1981) [English transl. in Math. USSR Sb. **43** (1982)]
8. Gonchar, A.A.: On the speed of rational approximation of some analytic functions. Mat. Sb. (NS) **105**, 147–163 (1978) [English transl. in Math. USSR Sb **34** (1978)]
9. Hernándes, B., De la Calli Ysern, B.: Meromorphic continuation of functions and arbitrary distribution of interpolation points. J. Math. Anal. Appl. **403**, 107–119 (2013)
10. Kakehashi, T.: The decomposition of coefficients of power series and the divergence of the interpolating polynomials. Proc. Jpn. Acad. **31**(8), 517–523 (1955)
11. Kakehashi, T.: On interpolation of analytic functions. II. Fundamental results. Proc. Jpn. Acad. **32**, 713–718 (1956)
12. Khristoforov, D.N.: On the asymptotic properties of interpolating polynomials. Matem. zametki **t.83**(1), 129–138 (2008) [English transl. in Math. Notes **83**(1), 116–124 (2008)]
13. Kovacheva, R.K.: Generalized Padé approximants and meromorphic continuation of functions. Mat. Sb **109** (1979) [English transl. Math. USSR Sb. **37**, 337–348 (1980)]
14. Landkoff, N.I.: Foundations of Modern Potential Theory, Grundlehren der mathematischen Wissenschaften. Springer, Berlin (1972)
15. Perron, O.: Die Lehre von den Kettenbrüchen, Vol. II, 3rd ed. Teubner Verlag, Stuttgart (1957)
16. Saff, E.B.: An extension of Montessus de Ballores theorem on the convergence of interpolating rational functions. J. Approx. Theory **6**, 63–67 (1972)
17. Saff, E.B., Totik, V.: Logarithmic Potentials with External Fields. Springer, Heidelberg (1997)
18. Suetin, S.P.: On the poles of the mth row of the Padé table. Mat. Sb. **120**(162), 500–504 (1983) [English transl. in Math. USSR Sb. 48 (1984)]
19. Suetin, S.P.: On an inverse problem for the mth row of the Padé table. Mat. Sb. **124**(166), 238–251 (1984) [English transl. in Math. USSR Sb. 52 (1985)]
20. Vavilov, V.V., López, G., Prokhorov, V.A.: On an inverse problem for the rows of the Padé table. Mat. Sb. 1**10**(152), 117–129 (1979) [English transl. in Math. USSR. 38 (1981)]
21. Walsh, J.L.: Overconvergence, degree of convergence and zeros of sequences of analytic functions. Duke Math. **13**, 195–234 (1946)

Total Phosphorus Removal in Horizontal Subsurface Flow Constructed Wetlands: A Computational Investigation for the Optimal Adsorption Model

Konstantinos Liolios, Vassilios Tsihrintzis, Panagiotis Angelidis, Krassimir Georgiev and Ivan Georgiev

Abstract A numerical simulation concerning Total Phosphorus (TP) removal in Horizontal Subsurface Flow Constructed Wetlands (HSF CWs) is presented. For the phenomenon of absorption, a comparison between the results of a linear and a non-linear model is realized. The purpose is to investigate which one of these two adsorption models is the optimal one for the computational simulation of TP removal. The simulations concern five pilot-scale HSF CWs units, which were constructed and operated in the facilities of the Laboratory of Ecological Engineering and Technology (LEET), Department of Environmental Engineering, Democritus University of Thrace (DUTh), Xanthi, Greece. Concerning the numerical simulation, the Visual MODFLOW computer code is used, which is based on the finite difference method. Finally, a comparison between computational and available experimental results is given.

Keywords Computational groundwater flow · Constructed wetlands · Total Phosphorus (TP) · Adsorption · MODFLOW · Freundlich and Langmuir Isotherms

K. Liolios (✉) · K. Georgiev · I. Georgiev
Institute of Information and Communication Technologies, Bulgarian Academy of Sciences, Sofia, Bulgaria
e-mail: kostisliolios@gmail.com

K. Georgiev
e-mail: georgiev@parallel.bas.bg

V. Tsihrintzis
Department of Infrastructure and Surveying Engineering, School of Rural and Surveying Engineering, National Technical University of Athens, Athens, Greece

P. Angelidis
Laboratory of Hydraulics and Hydraulic Works, Department of Civil Engineering, Democritus University of Thrace, Xanthi, Greece

I. Georgiev
Institute of Mathematics and Informatics, Bulgarian Academy of Sciences, Sofia, Bulgaria

K. Georgiev et al. (eds.), *Advanced Computing in Industrial Mathematics*, Studies in Computational Intelligence 728,
https://doi.org/10.1007/978-3-319-65530-7_11

1 Introduction

The use of Constructed Wetlands (CWs) is considered recently as a popular ecological and economical solution for the wastewater treatment, see e.g.[1–3]. The ability of these systems to remove pollutants is a very important factor during the optimal construction and operation of HSF CWs. The Total Phosphorus (TP) is one of the most common essential municipal pollutants, for which the phenomenon of adsorption requires special attention.

Many studies concerning the removal of TP in HSF CWs have been realized, especially in the previous decade. Most of them are experimental and the results describe the ability of various types of porous materials to remove TP in HSF CWs, see e.g. [4–14]. A few studies consider vertical flow (VF) CWs [15–17] or Free Water Surface (FWS) CWs [18, 19]. Relevant review studies, concerning the TP removal by using CWs, have been presented in [20, 21].

Recently, many computer codes for simulating the TP removal in CWs have been developed and successfully used. Especially for HSF CWs, some of the most representative models are CW2D [22, 23], CWM1 [24, 25] and BIO_PORE [26]. Similar computer codes have been developed for other types of CWs, e.g. FITOVERT [27] for VF CWs and DMSTA2 [28, 29] for FWS CWs. A detailed review for the available computer codes concerning the numerical modeling in CWs is given in [30–33].

The Artificial Neural Networks (ANN) procedure has been also successfully used for the prediction of the performance of HSF CWs for TP removal [34, 35]. More recently, a fuzzy logic model was used for describing the TP removal in FWS CWs [36].

In the present study, a numerical simulation concerning TP removal in HSF CW is presented. Emphasis is given to select the optimal adsorption model, by comparing computed results of procedures based either on the Freundlich (linear) or on the Langmuir (non-linear) isotherms. The Visual MODFLOW computer code [37], based on the Finite Difference Method (FDM), is used for the numerical simulations. The optimal values for the Langmuir parameters are estimated by using inverse problem procedures [38]. Finally, the computational results are compared with available experimental data, obtained from five pilot-scale HSF CWs, which were constructed and operated in LEET, DUTh, Xanthi, Greece.

2 The Mathematical Formulation of the Problem

The system of partial differential equations (PDE), which describes the advection, dispersion and removal of a solute in the three-dimensional (3-D) space, considering sources/sinks, equilibrated adsorption and first-order irreversible kinetic reactions, is in tensorial notation ($i, j = 1, 2, 3$) [39]:

$$\frac{\partial}{\partial x_i}\left(K_{ij}\frac{\partial h}{\partial x_j}\right)+\mathrm{q_v}=S_y\frac{\partial h}{\partial t} \tag{1}$$

$$v_i=-\frac{K_{ij}}{\theta}\frac{\partial h}{\partial x_j} \tag{2}$$

$$\theta R_d\frac{\partial C}{\partial t}=\frac{\partial}{\partial x_i}\left(\theta D_{ij}\frac{\partial C}{\partial x_j}\right)-\frac{\partial}{\partial x_i}(q_iC)+q_vC_s-\lambda_1\theta C-\lambda_2\rho_bS \tag{3}$$

The Eq. (1) concerns the groundwater flow and the Eq. (2) concerns the Darcy law in porous media.

In the above PDE, K_{ij} is a component of the hydraulic conductivity tensor, in [LT^{-1}]; h is the hydraulic head, in [L]; q_v is the volumetric flow rate per unit volume of aquifer, representing fluid sources (positive) or sinks (negative), in [T^{-1}]; S_y is the specific yield of the porous materials [dimensionless]; v_i is the seepage or linear pore water velocity, in [LT^{-1}], which is related to Darcy velocity q_i, in [LT^{-1}], through the relationship: $q_i = v_i\,\theta$; θ is the porosity [dimensionless]; R_d is the retardation factor [dimensionless]; C is the aqueous solute concentration, in [ML^{-3}]; D_{ij} is the hydrodynamic dispersion coefficient tensor, in [L^2T^{-1}]; C_s is the concentration of the source or sink flux, in [ML^{-3}]; ρ_b is the dry bulk density of the soil, in [ML^{-3}]; S is the concentration adsorbed by the solid phase of the porous medium, in [M pollutant/M solid]; λ_1 and λ_2 are the removal coefficients for the dissolved and adsorbed phases respectively, both in [T^{-1}]; Here, as usually in environmental engineering praxis, it is assumed: $\lambda_1 = \lambda_2 = \lambda$.

In Eq. (3), the retardation factor R_d depends on C and S, $R_d = R_d\,(C, S)$, and is given by the next equation:

$$R_d=1+\frac{\rho_b}{\theta}\frac{\partial S}{\partial C} \tag{4}$$

The values of bulk density ρ_b are calculated by the equation:

$$\rho_b=(1-\theta)\rho_r \tag{5}$$

where ρ_r is the density of solid grains of the porous material. Usually it is: ρ_r = 2.65 g/cm^3.

Regarding the dependence of S on C, i.e. $S = f(C)$, the most frequently used sorption isotherms are the Freundlich and Langmuir ones [39].

For a non-absorbable pollutant, e.g. Biochemical Oxygen Demand (BOD), it is S = 0 and R_d = 1. In this case, the Eq. (3) is a linear one.

On the contrary, for an absorbable pollutant, like TP, it is $S \neq 0$ and $R_d > 1$. In this case, the Eq. (3) can be a linear or a non-linear one, according to the dependence of S on C. The following cases are usually considered.

1. The non-linear Freundlich isotherm is expressed by the equation:

$$S = K_F C^{\alpha} \tag{6}$$

   where the constant K_F, in [L^3M^{-1}], and the exponent α [dimensionless] are experimentally estimated according to the type of pollutant and the porous medium. When $a \neq 1$, the Eq. (3) becomes non-linear.
2. For some pollutants, which have initial low concentrations, the phenomenon of adsorption is often described by the *linear isotherm of Freundlich*, where $\alpha = 1$. In this case, it is $K_F = K_d$ and the non-linear Freundlich isotherm is expressed by the equations:

$$S = K_d C \tag{7}$$

$$R_d = 1 + \frac{\rho_b}{\theta} K_d \tag{8}$$

   The parameter K_d is the distribution coefficient, in [L^3M^{-1}], which expresses the distribution of the pollutant concentrations between solid and liquid phases, S and C, respectively. For the case of the linear isotherm of Freundlich, Eq. (3) is a linear one.
3. The *non-linear Langmuir isotherm* is described by the equation:

$$S = S_{max} \frac{K_L C}{1 + K_L C} \tag{9}$$

   where S_{max} is the maximum adsorption capacity, in [M pollutant/M solid]; and K_L is the Langmuir constant, in [L^3M^{-1}]. Thus, for this case Eq. (3) is a high non-linear one.

   Taking into account Eqs. (6), (7) or (9), the above Eqs. (1)–(3), combined with appropriate boundary and initial conditions, formulate a system of Partial Differential Equations (PDE). The solution of this system provides the six main space-time functions of hydraulic head (h), Darcy velocities field (q_i) solute concentration (C) and adsorbed solid concentration (S): $h = h(x_i, t)$, $q_i = q_i(x_i, t)$, where $i = 1, 2, 3$, $C = C(x_i, t)$ and $S = S(x_i, t)$.

   The purpose of the present study is to investigate computationally which of the two adsorption models, the linear Freundlich isotherm of Eq. (7) or the non-linear Langmuir isotherm of Eq. (9), simulates better the experimental operation of the pilot-scale HSF CWs, described in [4, 40].

3 Numerical Simulation

For the numerical simulation of TP removal in the pilot-scale HSF CWs, the Visual MODFLOW family code [37], combined with the MT3DMS package [41], were used. MODFLOW is based on the Finite Difference Method (FDM) and is one of the most widely used computer codes for groundwater flow investigations. MT3DMS offers the possibility to simulate the removal of TP using both isotherm for adsorption, the Freundlich linear or the Langmuir non-linear one.

The values of the parameters λ and K_d of the Freundlich linear isotherm, see Eq. (7), have been determined and presented in [42]. The optimal values of the removal coefficient λ were assumed by adopting a trial-and-error procedure, similar to that described in [40]. The proper values for the distribution coefficient K_d were obtained according to the proposed range in the literature [39].

In this study, the simulation of the HSF CW's operation by using the non-linear Langmuir isotherm is presented. For this case, the values of the parameters K_L and S_{max}, see Eq. (9), should be estimated. First-order decay for TP removal was assumed and the same values of λ which were determined in [42] have been used, as the numerical simulation concerns the same facilities in both studies. The values for K_L and S_{max} were determined by using inverse problems procedures [38].

The computed TP concentration values were matched with the corresponding available experimental values. For verification, the model was run using the K_L and S_{max} values obtained in calibration, and a comparison was made between the computed and the experimental results of TP concentration at distances 1/3 and 2/3 of the CW unit length. As mentioned, concerning the linear case of the Freundlich isotherm, the results have already been presented in [42] and are used for the comparison and selection of the optimal adsorption model.

4 The Experimental Procedure

The experimental data, used for the calibration of the model and for the comparison between experimental and computational results, were collected from five pilot-scale HSF CWs. These units were constructed and operated for 2 years (2004–2006) in LEET, DUTh, Xanthi, Greece. A detailed description of these facilities has already been presented in [4, 40].

Briefly, the CW units had a rectangular scheme and contained various types of porous materials (i.e., medium gravel—MG, fine gravel—FG and cobbles—CO), and vegetation species (i.e., reed—R and cattails—C). One CW unit, denoted as Z, was kept unplanted for comparison reasons. The dimension of each tank was 3 m length, 0.75 m width and 1 m height, while the depth of porous material was 0.45 m.

5 Results and Discussion

5.1 Model Calibration—Determination of the Langmuir Parameters

As reported, available experimental data for five pilot-scale HSF CWs have been used for the values of the TP concentration at the inlet (C_{in}) and at the outlet (C_{out}) of the HSF CWs. The numerical simulation by using the Freundlich linear isotherm has been presented in [42], where the optimal values for λ and K_d have been estimated.

In the present study, where the simulation of TP removal by using the Langmuir non-linear isotherm is realized, the unknown parameters are the Langmuir constant K_L and the maximum adsorption capacity S_{max}, see Eq. (9). Considering λ, the values presented by [42] were reliable, as the experiments have been realized in the same facilities.

Based on the CW literature for similar porous media, see [6, 11], the range of the values for the unknowns parameters are: $K_L = 0.01$–0.10 L/mg and $S_{max} = 0.05$–0.20 mg TP/g soil. In order to determine the exact values of K_L, a trial and error procedure was adopted. The value of K_L, for which the computed values of the TP concentrations at the outlet were almost the same with the experimental ones, was $K_L = 0.01$ L/mg. By using this optimal value of K_L, inverse problems procedures were adopted [38] and more simulation tests were realized, in order to determine the values of S_{max}.

The simulation results, for each one of the five pilot-scale HSF CWs and for Hydraulic Residence Time (*HRT*) of 6, 8, 14 and 20 days, are presented in Table 1. This table provides, for $K_L = 0.01$ L/mg, representative values of the average temperature T_{av} [°C], the porosity θ [–], the bulk density ρ_b [g/cm^3], the inlet and outlet TP concentrations, C_{in} and C_{out} respectively [mg/L], the removal coefficient λ [d^{-1}] and the maximum adsorption capacity S_{max} [mg TP/g soil]. As these results show, the range of S_{max} was between 0.009 and 0.012 mg TP/g soil, also it is inside the range of other studies with similar operational and climatic conditions.

5.2 Verification of the Model—Comparison with Experimental Data

The next step, after the simulation, was to verify the model in order to check its accuracy. For this reason, a comparison between the computed and the experimental concentration values of TP was realized. These values concerned the concentration of TP among the length of HSF CWs, i.e. at distances 1/3 (1 m) and 2/3 (2 m) from the inlet of the pollutant, and were computationally estimated by using the λ, K_L and S_{max} values of Table 1.

Table 1 Estimated values of λ and S_{max}, for TP removal in the pilot-scale HSF CWs

HSF CW	*HRT* (days)	T_{av} (°C)	θ (–)	ρ_b (g/cm^3)	C_{in} (mg/L)	C_{out} (mg/L)	λ (d^{-1})	S_{max} (mg/g)
MG-R	6	15.7	0.35	1.723	9.2	9.2	0.000	0.009
	8	12.1	0.36	1.709	9.2	7.5	0.015	0.012
	14	16.2	0.38	1.643	9.7	5.8	0.023	0.011
	20	15.2	0.35	1.723	8.8	5.7	0.012	0.010
MG-C	6	15.7	0.33	1.776	9.2	7.0	0.026	0.011
	8	12.1	0.34	1.762	9.2	3.9	0.062	0.011
	14	16.2	0.34	1.749	9.7	2.3	0.061	0.010
	20	15.2	0.33	1.776	8.8	1.8	0.046	0.011
MG-Z	6	15.7	0.37	1.670	9.2	5.8	0.047	0.010
	8	12.1	0.37	1.670	9.2	5.9	0.034	0.011
	14	16.2	0.37	1.670	9.7	5.5	0.025	0.011
	20	15.2	0.37	1.670	8.8	4.5	0.020	0.011
FG-R	6	15.7	0.29	1.882	9.2	2.8	0.107	0.010
	8	12.1	0.29	1.882	9.2	1.4	0.128	0.011
	14	16.2	0.31	1.829	9.7	1.1	0.090	0.010
	20	15.2	0.29	1.882	8.8	1.1	0.056	0.010
CO-R	6	15.7	0.28	1.908	9.2	6.9	0.025	0.012
	8	12.1	0.28	1.908	9.2	5.5	0.025	0.011
	14	16.2	0.28	1.908	9.7	1.6	0.069	0.010
	20	15.2	0.28	1.908	8.8	2.5	0.034	0.010

Two mathematical criteria were used. First, linear regression lines of the form $y = \gamma x$. Second, the coefficient of determination (R^2). Both criteria show the connection between computed (y) and experimental (x) values. For best fit, both parameters, the slope γ and R^2, should be as close to 1.0 as possible. The results, for each one of the five HSF CW pilot-scale units, are presented in Figs. 1, 2, 3, 4 and 5.

According to the first criterion, the slope γ is slightly higher than 1.0 for the MG-R and MG-Z units. This shows that the model overestimates the TP concentration values for these tanks. For the other three HSF CW units (MG-C, FG-R and CO-R) the values of slope γ are lower than 1.0. Generally, all the γ values are close enough to 1.0 and this indicates that the model describes satisfactory the phenomenon. Regarding the coefficient of determination R^2, its values were higher than 0.50 for most HSF CW units, and in most cases are satisfactorily close to 1.0, especially in comparison with other similar studies. These results show that the model can describe with good accuracy the physical phenomenon of TP removal in HSF CWs.

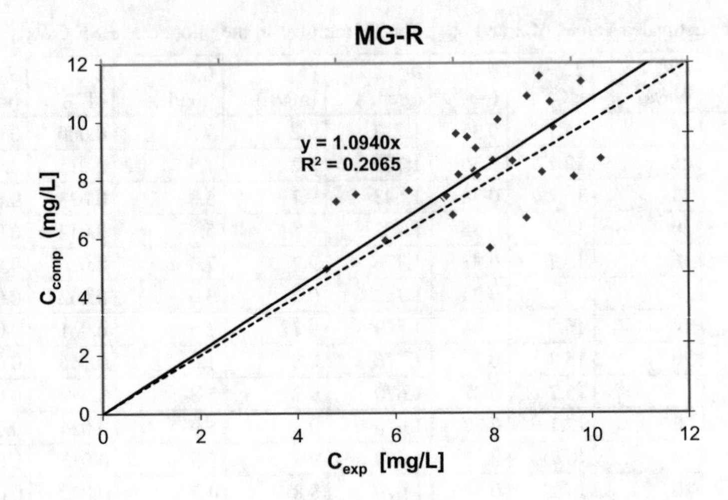

Fig. 1 Verification of the Langmuir adsorption model for the MG-R unit

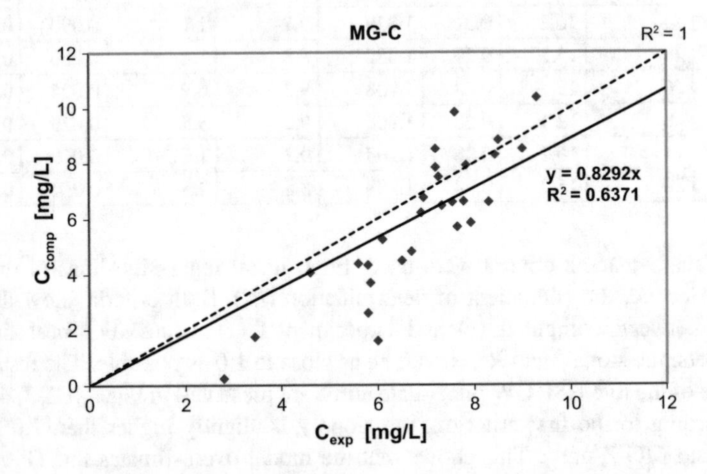

Fig. 2 Verification of the Langmuir adsorption model for the MG-C unit

6 Selection of the Optimal Adsorption Model: Linear or Non-linear?

The main purpose of this study was to decide which one of the two most well-known adsorption models, the linear Freundlich or the non-linear Langmuir, was the optimal one for the numerical simulation of the TP removal in the DUTh pilot-scale HSF CWs.

In the previous Section, the verification of the model concerning the Langmuir isotherm was presented and values for the slope γ, of the linear regression lines of

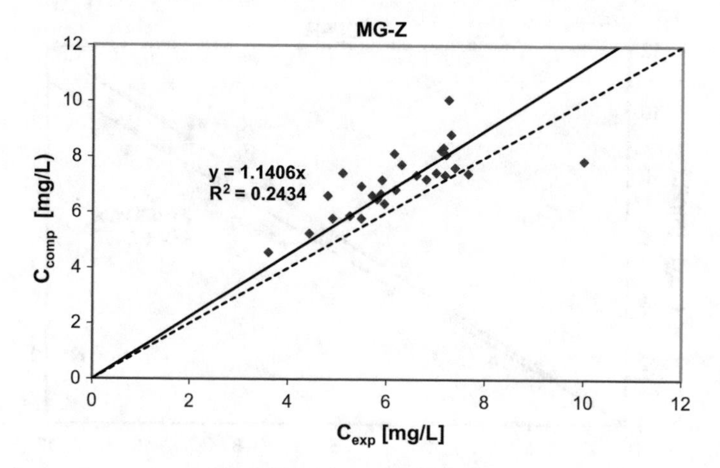

Fig. 3 Verification of the Langmuir adsorption model for the MG-Z unit

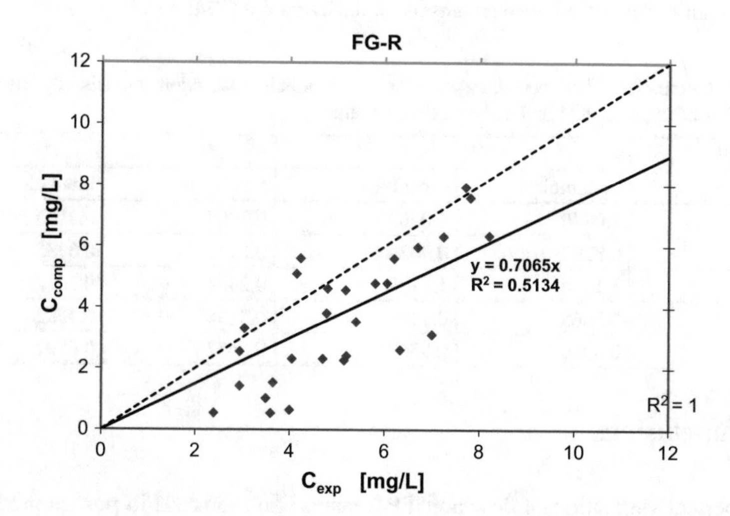

Fig. 4 Verification of the Langmuir adsorption model for the FG-R unit

the form $y = \gamma x$, and for the coefficient of determination (R^2) were determined. In order to select the optimal adsorption model, a comparison of these values (γ and R^2) which were determined in the present study for Langmuir isotherm and in [42] for Freundlich isotherm, has been realized. The results are presented in next Table 2.

As the results of Table 2 show, the Freundlich linear adsorption model describes slightly better the removal of TP in the pilot-scale HSF CWs, comparing with the Langmuir non-linear one.

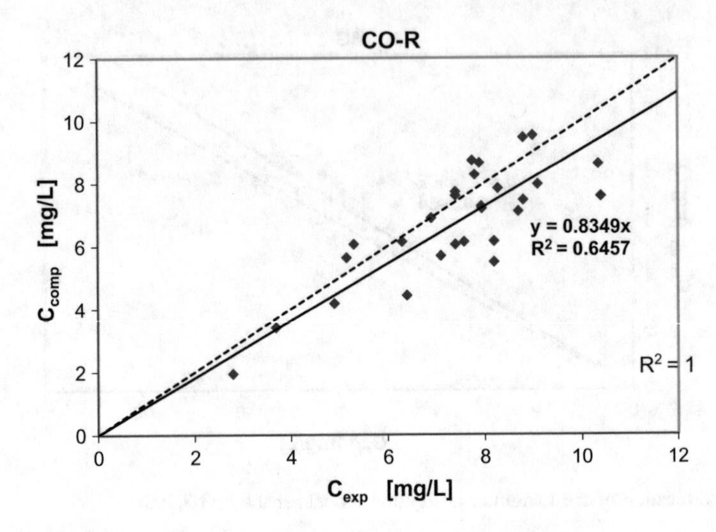

Fig. 5 Verification of the Langmuir adsorption model for the CO-R unit

Table 2 Comparison between Langmuir and Freundlich adsorption models, by using the correlation of experimental and computational results

HSF CW	γ		R^2	
	Langmuir	Freundlich	Langmuir	Freundlich
MG-R	1.0940	1.0562	0.2065	0.2166
MG-C	0.8292	0.9024	0.6371	0.6435
MG-Z	1.1406	1.1188	0.2434	0.2612
FG-R	0.7065	0.7494	0.5134	0.5592
CO-R	0.8349	0.9369	0.6457	0.6597

7 Conclusions

A numerical simulation of flow and TP transport and removal in porous media has been presented. Emphasis was given to select the optimal adsorption model, concerning the available experimental results of five pilot-scale HSF CW units. For this purpose, a computational investigation was realized, by using the Visual MODFLOW computer code. For the Langmuir non-isotherm adsorption model, the main unknown parameters λ, S_{max} and K_L were estimated, by using inverse methods. The results show that the linear Freundlich isotherm describes slightly better the operation for the pilot-scale HSF CW units, in comparison to the non-linear Langmuir isotherm. The proposed values of the parameters for both adsorption models (λ, K_d, K_L and S_{max}) can be used effectively for the optimum design of HSF CWs, both pilot-scale or full-scale, which have operation similarities with the DUTh units. Moreover, for such new facilities, their construction and operation could be realized in the best ecological and economical way.

Acknowledgements This work was supported by the Bulgarian Academy of Sciences through the "Young Scientists" Grant No. DFNP-97/04.05.2016.

References

1. Kadlec, R.H., Wallace, S.D.: Treatment Wetlands, 2nd edn. CRC Press, Boca Raton (2009)
2. Vymazal, J., Kröpfelová, L.: Wastewater Treatment in Constructed Wetlands with Horizontal Sub-Surface Flow. Springer, Berlin (2008)
3. Gikas, G.D., Tsihrintzis, V.A.: On-site treatment of domestic wastewater using a small-scale horizontal subsurface flow constructed wetland. Water Sci. Technol. **62**(3), 603–614 (2010)
4. Akratos, C.S., Tsihrintzis, V.A.: Effect of temperature, HRT, vegetation and porous media on removal efficiency of pilot-scale horizontal subsurface flow constructed wetlands. Ecol. Eng. **29**(2), 173–191 (2007)
5. Vymazal, J.: Removal of phosphorus in constructed wetlands with horizontal sub-surface flow in the Czech Republic. Water Air Soil Pollut. Focus **4**(2), 657–670 (2004)
6. Drizo, A., Frost, C.A., Grace, J., Smith, K.A.: Phosphate and ammonium distribution in a pilot-scale constructed wetland with horizontal subsurface flow using shake as a substrate. Water Res. **34**(9), 2483–2490 (2000)
7. Brix, H., Arias, C.A., Del Bubba, M.: Media selection for sustainable phosphorus removal in subsurface flow constructed wetlands. Water Sci. Technol. **44**(11–12), 47–54 (2001)
8. Huett, D.O., Morris, S.G., Smith, G., Hunt, N.: Nitrogen and phosphorus removal from plant nursery runoff in vegetated and unvegetated subsurface flow wetlands. Water Res. **39**(14), 3259–3272 (2005)
9. Arias, C.A., Del Bubba, M., Brix, M.: Phosphorus removal by sands for use as media in subsurface flow constructed wetlands. Water Res. **35**(5), 1159–1168 (2001)
10. Molle, P., Martion, S., Esser, D., Besnault, S., Morlay, C., Harouiya, N.: Phosphorus removal by use of apatite in constructed wetlands: Design recommendations. Water Pract. Technol. **6**(3), wpt2011046 (2011)
11. Del Bubba, M., Arias, C.A., Brix, M.: Phosphorus adsorption maximum of sands for use as media in subsurface flow constructed reed beds as measured by the Langmuir isotherm. Water Res. **37**(14), 3390–3400 (2003)
12. Cui, L., Zhu, X., Ma, M., Ouyang, Y., Dong, M., Zhu, W., Luo, S.: Phosphorus sorption capacities and physicochemical properties of nine substrate materials for constructed wetlands. Arch. Environ. Con. Tox. **55**(2), 210–217 (2008)
13. Pant, H.K., Reddy, K.R., Lemon, E.: Phosphorus retention capacity of root bed media of sub-surface flow constructed wetlands. Ecol. Eng. **17**(4), 345–355 (2001)
14. Vohla, C., Alas, R., Nurk, K., Baatz, S., Mander, Ü.: Dynamics of phosphorus, nitrogen and carbon removal in a horizontal subsurface flow constructed wetland. Sci. Total Environ. **380** (1–5), 66–74 (2007)
15. Prochaska, C.A., Zouboulis, A.I.: Removal of phosphates by pilot vertical-flow constructed wetlands using a mixture of sand and dolomite as substrate. Ecol. Eng. **26**(3), 293–303 (2006)
16. Stefanakis, A.I., Tsihrintzis, V.A.: Performance of pilot-scale vertical flow constructed wetlands treating simulated municipal wastewater: effect of various design parameters. Desalination **248**(1–3), 753–770 (2009)
17. Stefanakis, A., Akratos, C.S., Tsihrintzis, V.A.: Vertical Flow Constructed Wetlands: Eco-Engineering Systems for Wastewater and Sludge Treatment. Elsevier, Amsterdam (2014)
18. Kulabako, N.R., Nalubega, M., Thunvik, R.: Phosphorus transport in shallow groundwater in peri-urban Kambala, Uganda: results from field and laboratory measurements. Environ. Geol. **53**(7), 1535–1551 (2008)

19. Jamieson, R., Gordon, R., Wheeler, N., Smith, E., Stratton, G., Madani, A.: Determination of first order rate constants for wetlands treating livestock wastewater in cold climates. J. Environ. Eng. Sci. **6**(1), 65–72 (2007)
20. Reddy, K.R., Kadlec, R.H., Flaig, E., Gale, P.M.: Phosphorus retention in streams and wetlands: A review. Crit. Rev. Env. Sci. Tech. **29**(1), 83–146 (1999)
21. Vohla, C., Koiv, M., Bavor, H.J., Chazarenc, F., Mander, Ü.: Filter materials for phosphorus removal from wastewater in treatment wetlands: a review. Ecol. Eng. **37**(1), 70–89 (2001)
22. Langergraber, G., Simunek, J.: Modeling variably saturated water flow and multicomponent reactive transport in constructed wetlands. Vadoze Zone J. **4**(4), 924–938 (2005)
23. Toscano, A., Langergraber, G., Consoli, S., Cirelli, G.L.: Modelling pollutant removal in a pilot-scale two-stage subsurface flow constructed wetland. Ecol. Eng. **35**(2), 281–289 (2009)
24. Langergraber, G., Rousseau, D.P.L., Garcia, J., Mena, J.: CWM1: A general model to describe biokinetic processes in subsurface flow constructed wetlands. Water Sci. Technol. **59** (9), 1687–1697 (2009)
25. Llorens, E., Saaltink, M.W., Garcia, J.: CWM1 implementation in RetrasoCodeBright: First results using horizontal subsurface flow constructed wetland data. Chem. Eng. J. **166**(1), 224–232 (2011)
26. Samso, R., Garcia, J.: BIO_PORE, a mathematical model to simulate biofilm growth and water quality improvement in porous media: Application and calibration for constructed wetlands. Ecol. Eng. **54**, 116–127 (2013)
27. Giraldi, D., de Michieli, V.M., Iannelli, R.: FITOVERT: A dynamic numerical model of subsurface vertical flow constructed wetlands. Environ. Model. Soft. **25**(5), 633–640 (2010)
28. Walker, W.W.J., Kadlec, R.H.: Modeling phosphorus dynamics in Everglades wetlands and stormwater treatment areas. Crit. Rev. Environ. Sci. Tech. **41**, 430–446 (2011)
29. Wang, Y.C., Lin, Y.P., Huang, C.W., Chiang, L.C., Chu, H.J., Ou, W.S.: A system dynamic model and sensitivity analysis for simulating domestic pollution removal in a free-water surface constructed wetland. Water Air Soil Pollut. **223**(5), 2719–2742 (2012)
30. Kumar, J.L.G., Zhao, Y.Q.: A review on numerous modeling approaches for effective, economical and ecological treatment wetlands. J. Environ. Manag. **92**(3), 400–406 (2011)
31. Langergraber, G., Giraldi, D., Mena, J., Meyer, D., Pena, M., Toscano, A., Brovelli, A., Korkusuz, E.A.: Recent developments in numerical modelling of subsurface flow constructed wetlands. Sci. Total Environ. **407**(13), 3931–3943 (2009)
32. Rousseau, D.P.L., Vanrolleghem, P.A., De Pauw, N.: Model-based design of horizontal subsurface flow constructed treatment wetlands: a review. Water Res. **38**(6), 1484–1493 (2004)
33. Lewis, D.R., McGechan, M.B.: A review of field scale phosphorus dynamics models. Biosyst. Eng. **82**(4), 359–380 (2002)
34. Akratos, C.S., Papaspyros, J.N.E., Tsihrintzis, V.A.: Artificial neural network use in ortho-phosphate and total phosphorus removal prediction in horizontal subsurface flow constructed wetlands. Biosyst. Eng. **102**(2), 190–201 (2009)
35. Song, K., Park, Y.S., Zheng, F., Kang, H.: The application of artificial neural network (ANN) model to the simulation of denitrification rates in mesocosm-scale wetlands. Ecol. Inform. **16**, 10–16 (2013)
36. Kotti, I.P., Sylaios, G.K., Tsihrintzis, V.A.: Fuzzy modeling for nitrogen and phosphorus removal estimation in free-water surface constructed wetlands. Environ. Proc. **3**(1), 65–79 (2016)
37. Waterloo Hydrogeologic Inc.: Visual MODFLOW v. 4.2. User's Manual. U.S. Geological Survey, Virginia (2006)
38. Sun, N.Z.: Inverse Problems in Groundwater Modeling. Springer, Berlin (2013)
39. Zheng, C., Bennett, G.D.: Applied Contaminant Transport Modelling, 2nd edn. Wiley, New York (2002)
40. Liolios, K.A., Moutsopoulos, K.N., Tsihrintzis, V.A.: Modeling of flow and BOD fate in horizontal subsurface flow constructed wetlands. Chem. Eng. J. **200–202**, 681–693 (2012)

41. Zheng, C.: MT3D: A modular three-dimensional transport model for simulation of advection, dispersion and chemical reactions of contaminants in groundwater systems. Report to the U.S. Environmental Protection Agency, University of Alabama, Tuscaloosa (1990)
42. Liolios, K.A., Moutsopoulos, K.N., Tsihrintzis, V.A.: Numerical simulation of phosphorus removal in horizontal subsurface flow constructed wetlands. Desal. Water Treat. **56**(5), 1282–1290 (2015)

Some Results Involving Euler-Type Integrals and Dilogarithm Values

Lubomir Markov

Abstract The claim that $\mathrm{Li}_2\left(-\frac{\sqrt{5}+1}{2}\right) = -\frac{\pi^2}{10} + \frac{1}{2}\log^2\left(\frac{\sqrt{5}+1}{2}\right)$, circulating in the dilogarithm literature since at least 1958, is wrong. We derive the correct value $\mathrm{Li}_2\left(-\frac{\sqrt{5}+1}{2}\right) = -\frac{\pi^2}{10} - \log^2\left(\frac{\sqrt{5}+1}{2}\right)$ and use it to obtain several formulas for π^2 in terms of dilogarithm values at the "golden relatives" $\frac{1}{\phi^2}, \frac{1}{\phi}, -\frac{1}{\phi}, -\phi$ of $\phi = \frac{\sqrt{5}+1}{2}$. We also sum the series $\sum_{n=0}^{\infty} \frac{G_N(n)}{(2n+1)^3}$ and $\sum_{n=1}^{\infty} \frac{H_N(n)}{n^3}$ in terms of Euler-type integrals $\int_0^{\frac{\pi}{2}} x^M \log(\sin x)\,\mathrm{d}x$, where $G_N(n)$ and $H_N(n)$ are the quantities appearing in the Borwein-Chamberland expansions of $\arcsin^{2N+1}(z)$ and $\arcsin^{2N}(z)$, respectively. As special cases we obtain very simple proofs of Euler's equation

$$\zeta(3) = \frac{2\pi^2}{7}\log 2 + \frac{16}{7}\int_0^{\frac{\pi}{2}} x \log(\sin x)\mathrm{d}x$$

and of the similar formula

$$\zeta(3) = \frac{2\pi^2}{9}\log 2 + \frac{16}{3\pi}\int_0^{\frac{\pi}{2}} x^2 \log(\sin x)\mathrm{d}x.$$

Keywords Dilogarithm · Golden ratio · Riemann zeta function · Euler-type integrals · Borwein-Chamberland expansions · Euler's equation for $\zeta(3)$ · Wallis integrals

---

L. Markov (✉)
Department of Mathematics and CS, Barry University,
11300 N.E. Second Avenue, Miami Shores, FL 33161, USA
e-mail: lmarkov@barry.edu

K. Georgiev et al. (eds.), *Advanced Computing in Industrial Mathematics*, Studies in Computational Intelligence 728,
https://doi.org/10.1007/978-3-319-65530-7_12

1 Introduction

Let $\zeta(z)$ be the Riemann zeta function and $\mathrm{Li}_2(z)$ be the dilogarithm. Both functions occupy a prominent place in mathematics, and have increasingly appeared in modern physics (see for example [6, 9, 12]).

At present, the evaluation of the dilogarithm in closed form is known to be possible only at eight values for z, the first four being the rational numbers $0, 1, -1$ and $\frac{1}{2}$. The remaining four are the irrational numbers $\frac{1}{\phi^2}, \frac{1}{\phi}, -\frac{1}{\phi}$ and $-\phi$, where $\phi = \frac{\sqrt{5}+1}{2}$ is the golden ratio. The evaluation of Li_2 at the first seven numbers has been performed by Landen [4] and is sometimes called "Landen's list". We turn our attention to $\mathrm{Li}_2(-\phi)$. It seems inconceivable that the value for it circulating in the literature (see [5, 6, 8, 10, 12]) since at least 1958, namely $\mathrm{Li}_2\left(-\frac{\sqrt{5}+1}{2}\right) = -\frac{\pi^2}{10} + \frac{1}{2}\log^2\left(\frac{\sqrt{5}+1}{2}\right)$, is wrong. The correct formula is $\mathrm{Li}_2\left(-\frac{\sqrt{5}+1}{2}\right) = -\frac{\pi^2}{10} - \log^2\left(\frac{\sqrt{5}+1}{2}\right)$, which we derive below. (We do so without claim to priority; see Sect. 4.) With regard to this astonishingly resilient error, one might find refreshing the following quotation from Zagier [12] (who also gives the incorrect value):

> First defined by Euler, it [the dilogarithm] has been studied by some of the great mathematicians of the past - Abel, Lobachevsky, Kummer, and Ramanujan ... Almost all of its appearances in mathematics, and almost all the formulas relating to it, have something of the fantastical in them, as if this function alone among all others possessed a sense of humor.

As concerns the evaluation of $\zeta(z)$, it is well-known that $\zeta(2) = \frac{\pi^2}{6}$, $\zeta(4) = \frac{\pi^4}{90}$, and in general $\zeta(2n) = \frac{(-1)^{n+1}B_{2n}}{2(2n)!}(2\pi)^{2n}$, where B_{2n} are the Bernoulli numbers. No such formula is known for the values of the zeta function at odd integers, and it is an open problem of the first importance whether $\zeta(2n+1)$ is expressible in terms of known constants.

In a classical paper [3], Euler derives the following representation for $\zeta(3)$ which he obviously considers of significant interest:

$$\zeta(3) = \frac{2\pi^2}{7}\log 2 + \frac{16}{7}\int_0^{\frac{\pi}{2}} x\log(\sin x)\mathrm{d}x \tag{1}$$

His proof uses intricate manipulations of divergent and convergent series and takes up 16 pages. Our aim is to give a very simple proof (we believe the simplest so far) of this formula, obtained as a special case of a more general result (Theorem 1 below). Another interesting representation implied by Theorem 1 is

$$\zeta(3) = \frac{2\pi^2}{9}\log 2 + \frac{16}{3\pi}\int_0^{\frac{\pi}{2}} x^2 \log(\sin x)\mathrm{d}x. \tag{2}$$

The integrals in the last two formulas may be termed Euler-type integrals. Formulas involving $\zeta(3)$ in terms of similar integrals have recently been found to be of importance in physics (see [10]).

2 $\mathrm{Li}_2(-\phi)$ and Some Related Formulas

For the correct calculation of $\mathrm{Li}_2(-\phi)$ (cf. [5, pp. 1–7]), recall that one first establishes $\mathrm{Li}_2\Big(\frac{\sqrt{5}-1}{2}\Big) = \frac{\pi^2}{10} - \log^2\Big(\frac{\sqrt{5}-1}{2}\Big)$. In Landen's functional equation

$$\mathrm{Li}_2(x) + \mathrm{Li}_2\Big(\frac{x}{x-1}\Big) = -\frac{1}{2}\log^2(1-x),$$

the argument x is taken to be $\frac{\sqrt{5}-1}{2}$. This yields

$$\mathrm{Li}_2\Big(\frac{\sqrt{5}-1}{2}\Big) + \mathrm{Li}_2\Big(-\frac{\sqrt{5}+1}{2}\Big) = -\frac{1}{2}\log^2\Big(\frac{3-\sqrt{5}}{2}\Big).$$

Substituting $\mathrm{Li}_2\Big(\frac{\sqrt{5}-1}{2}\Big)$, using the relations $\frac{\sqrt{5}-1}{2} = \frac{2}{\sqrt{5}+1}$, $\frac{3-\sqrt{5}}{2} = \Big(\frac{2}{\sqrt{5}+1}\Big)^2$ and simplifying gives the correct value:

$$\mathrm{Li}_2\Big(-\frac{\sqrt{5}+1}{2}\Big) = -\frac{\pi^2}{10} - \log^2\Big(\frac{\sqrt{5}+1}{2}\Big). \tag{3}$$

Let us call $\frac{1}{\phi^2}$, $\frac{1}{\phi}$, $-\frac{1}{\phi}$, $-\phi$ the "golden relatives" of ϕ. The second half of the expanded Landen's list (*cf.* [5]) becomes:

$$\begin{aligned}
\mathrm{Li}_2\Big(\frac{3-\sqrt{5}}{2}\Big) &= \mathrm{Li}_2\Big(\frac{1}{\phi^2}\Big) = \frac{\pi^2}{15} - \log^2(\phi),\\
\mathrm{Li}_2\Big(\frac{\sqrt{5}-1}{2}\Big) &= \mathrm{Li}_2\Big(\frac{1}{\phi}\Big) = \frac{\pi^2}{10} - \log^2(\phi),\\
\mathrm{Li}_2\Big(\frac{1-\sqrt{5}}{2}\Big) &= \mathrm{Li}_2\Big(-\frac{1}{\phi}\Big) = -\frac{\pi^2}{15} + \frac{1}{2}\log^2(\phi),\\
\mathrm{Li}_2\Big(-\frac{\sqrt{5}+1}{2}\Big) &= \mathrm{Li}_2(-\phi) = -\frac{\pi^2}{10} - \log^2(\phi).
\end{aligned}$$

Some interesting formulas for π^2 in terms of dilogarithm values at the golden relatives follow easily from the above; the following three relations:

$$\mathrm{Li}_2\Big(\frac{1}{\phi^2}\Big)+\mathrm{Li}_2\Big(\frac{1}{\phi}\Big)+2\mathrm{Li}_2\Big(-\frac{1}{\phi}\Big)-\mathrm{Li}_2(-\phi)=\frac{2\pi^2}{15},$$
$$\mathrm{Li}_2\Big(\frac{1}{\phi^2}\Big)-\mathrm{Li}_2\Big(\frac{1}{\phi}\Big)-2\mathrm{Li}_2\Big(-\frac{1}{\phi}\Big)-\mathrm{Li}_2(-\phi)=\frac{3\pi^2}{15}=\frac{\pi^2}{5},$$
$$-\mathrm{Li}_2\Big(\frac{1}{\phi^2}\Big)+\mathrm{Li}_2\Big(\frac{1}{\phi}\Big)-2\mathrm{Li}_2\Big(-\frac{1}{\phi}\Big)-\mathrm{Li}_2(-\phi)=\frac{4\pi^2}{15},$$

together with the especially pleasing formula:

$$2\mathrm{Li}_2\Big(\frac{1}{\phi^2}\Big)-2\mathrm{Li}_2\Big(\frac{1}{\phi}\Big)-2\mathrm{Li}_2\Big(-\frac{1}{\phi}\Big)-\mathrm{Li}_2(-\phi)=\frac{\pi^2}{6}=\mathrm{Li}_2(1),$$

are directly verified.

On the other hand, upon eliminating π^2– terms, one obtains:

$$2\mathrm{Li}_2\Big(\frac{1}{\phi^2}\Big)-\mathrm{Li}_2\Big(\frac{1}{\phi}\Big)+2\mathrm{Li}_2\Big(-\frac{1}{\phi}\Big)-\mathrm{Li}_2(-\phi)=\log^2(\phi).$$

3 Two Series in Terms of Euler-Type Integrals

In this section, we sum two interesting series in terms of Euler-type integrals $\int_0^{\frac{\pi}{2}} x^M\cdot\log(\sin x)\,\mathrm{d}x$, $M=0,1,2,\ldots$. Recall that $\int_0^{\frac{\pi}{2}}\log(\sin x)\,\mathrm{d}x=-\frac{\pi}{2}\log 2$. For $M\geq 1$, the integrals $\int_0^{\frac{\pi}{2}} x^M\log(\sin x)\,\mathrm{d}x$ have not been evaluated, at present, in terms of known constants.

Our result depends on the series expansions of integer powers of arcsin(z), which were discovered recently by Borwein and Chamberland [1]. We observe that the coefficients in these expansions involve quantities that are reciprocals of values of Wallis integrals (*cf.* [2]).

For odd powers of $f(z)=\arcsin(z)$, one has:

$$\arcsin(z)=\sum_{n=0}^{\infty}\frac{\binom{2n}{n}}{(2n+1)4^n}z^{2n+1},$$
$$\arcsin^3(z)=6\sum_{n=0}^{\infty}\Big\{\sum_{m=0}^{n-1}\frac{1}{(2m+1)^2}\Big\}\frac{\binom{2n}{n}}{(2n+1)4^n}z^{2n+1},$$
$$\arcsin^5(z)=120\sum_{n=0}^{\infty}\Big\{\sum_{m=0}^{n-1}\frac{1}{(2m+1)^2}\sum_{p=0}^{m-1}\frac{1}{(2p+1)^2}\Big\}\frac{\binom{2n}{n}}{(2n+1)4^n}z^{2n+1}.$$

The general formula is

$$\arcsin^{2N+1}(z) = (2N+1)! \sum_{n=0}^{\infty} \frac{G_N(n)\binom{2n}{n}}{(2n+1)4^n} z^{2n+1}, \tag{4}$$

where $G_0(n) = 1$, and

$$G_N(n) = \sum_{m_1=0}^{n-1} \frac{1}{(2m_1+1)^2} \sum_{m_2=0}^{m_1-1} \frac{1}{(2m_2+1)^2} \cdots \sum_{m_N=0}^{m_{N-1}-1} \frac{1}{(2m_N+1)^2}.$$

The expansions for the even powers are:

$$\arcsin^2(z) = \frac{1}{2} \sum_{n=1}^{\infty} \frac{4^n}{n^2\binom{2n}{n}} z^{2n},$$

$$\arcsin^4(z) = \frac{3}{2} \sum_{n=1}^{\infty} \left\{ \sum_{m=1}^{n-1} \frac{1}{m^2} \right\} \frac{4^n}{n^2\binom{2n}{n}} z^{2n},$$

$$\arcsin^6(z) = \frac{45}{4} \sum_{n=1}^{\infty} \left\{ \sum_{m=1}^{n-1} \frac{1}{m^2} \sum_{p=1}^{m-1} \frac{1}{p^2} \right\} \frac{4^n}{n^2\binom{2n}{n}} z^{2n}.$$

In general,

$$\arcsin^{2N}(z) = (2N)! \sum_{n=1}^{\infty} \frac{H_N(n)\, 4^n}{n^2\binom{2n}{n}} z^{2n}, \tag{5}$$

where $H_1(n) = \frac{1}{4}$, and

$$H_{N+1}(n) = \frac{1}{4} \sum_{m_1=1}^{n-1} \frac{1}{(2m_1)^2} \sum_{m_2=1}^{m_1-1} \frac{1}{(2m_2)^2} \cdots \sum_{m_N=1}^{m_{N-1}-1} \frac{1}{(2m_N)^2}.$$

Theorem 1 *There hold the relations*

$$(A) \sum_{n=0}^{\infty} \frac{G_N(n)}{(2n+1)^3} = -\frac{\pi}{2(2N)!} \int_0^{\frac{\pi}{2}} x^{2N} \log(\sin x)\, dx + \frac{2N+2}{(2N+1)!} \int_0^{\frac{\pi}{2}} x^{2N+1} \log(\sin x)\, dx,$$

$$(B) \sum_{n=1}^{\infty} \frac{H_N(n)}{n^3} = -\frac{2}{(2N-1)!} \int_0^{\frac{\pi}{2}} x^{2N-1} \log(\sin x)\, dx + \frac{4(2N+1)}{\pi(2N)!} \int_0^{\frac{\pi}{2}} x^{2N} \log(\sin x)\, dx,$$

for $N = 0, 1, 2, \ldots$ in (A) and $N = 1, 2, 3, \ldots$ in (B).

Setting $N = 0$ in (A) yields Euler's formula (1). Setting $N = 1$ in (B) and combining with (1) gives Eq. (2).

Proof (A): Take Eq. (4), divide both sides by z, and integrate from 0 to x:

$$\int_0^x \frac{\arcsin^{2N+1}(z)}{z} dz = (2N+1)! \sum_{n=0}^{\infty} \frac{G_N(n)}{(2n+1)^2} \frac{\binom{2n}{n}}{4^n} x^{2n+1};$$

put $x = \sin\theta$ and integrate from 0 to $\dfrac{\pi}{2}$:

$$\int_0^{\frac{\pi}{2}}\int_0^{\sin\theta}\frac{\arcsin^{2N+1}(z)}{z}\mathrm{d}z\,\mathrm{d}\theta = (2N+1)!\sum_{n=0}^{\infty}\frac{G_N(n)}{(2n+1)^2}\frac{\binom{2n}{n}}{4^n}\int_0^{\frac{\pi}{2}}(\sin\theta)^{2n+1}\,\mathrm{d}\theta.$$

(The interchanging of summation and integration is justified by uniform convergence.)

On the left-hand side we have:

$$\int_0^{\frac{\pi}{2}}\int_0^{\sin\theta}\frac{\arcsin^{2N+1}(z)}{z}\,\mathrm{d}z\,\mathrm{d}\theta = \int_0^1\int_{\arcsin z}^{\frac{\pi}{2}}\frac{\arcsin^{2N+1}(z)}{z}\,\mathrm{d}\theta\,\mathrm{d}z$$
$$= \int_0^1\frac{\arcsin^{2N+1}(z)}{z}\left[\frac{\pi}{2}-\arcsin z\right]\mathrm{d}z = \frac{\pi}{2}\int_0^1\frac{\arcsin^{2N+1}(z)}{z}\,\mathrm{d}z$$
$$-\int_0^1\frac{\arcsin^{2N+2}(z)}{z}\,\mathrm{d}z$$
$$= -\frac{\pi}{2}(2N+1)\int_0^{\frac{\pi}{2}}x^{2N}\log(\sin x)\,\mathrm{d}x + (2N+2)\int_0^{\frac{\pi}{2}}x^{2N+1}\log(\sin x)\,\mathrm{d}x.$$

On the right-hand side, the Wallis integral $\int_0^{\frac{\pi}{2}}(\sin\theta)^{2n+1}\,\mathrm{d}\theta$ equals $\frac{(2n)!!}{(2n+1)!!}$, and we obtain

$$(2N+1)!\sum_{n=0}^{\infty}\frac{G_N(n)}{(2n+1)^2}\frac{\binom{2n}{n}}{4^n}\int_0^{\frac{\pi}{2}}(\sin\theta)^{2n+1}\,\mathrm{d}\theta =$$
$$(2N+1)!\sum_{n=0}^{\infty}\frac{G_N(n)}{(2n+1)^2}\frac{\binom{2n}{n}}{4^n}\frac{(2n)!!}{(2n+1)!!} = (2N+1)!\sum_{n=0}^{\infty}\frac{G_N(n)}{(2n+1)^3}.$$

This proves (A).

For the proof of (B), begin with Eq. (5) and repeat the same procedure as above. This time one needs the Wallis integral $\int_0^{\frac{\pi}{2}}(\sin\theta)^{2n}\,\mathrm{d}\theta = \frac{\pi}{2}\frac{(2n-1)!!}{(2n)!!}$, where $(-1)!!$ is defined to be 1. All steps are similar and we leave their verification to the reader.

4 Conclusions and Final Remarks

In this work we proved that the correct value for $\mathrm{Li}_2(-\phi)$ is $\mathrm{Li}_2\left(-\frac{\sqrt{5}+1}{2}\right) = -\frac{\pi^2}{10}-\log^2\left(\frac{\sqrt{5}+1}{2}\right)$. We then derived some formulas for π^2 and $\log^2(\phi)$ in terms of dilogarithm values at the "golden relatives" of the golden number $\phi = \frac{\sqrt{5}+1}{2}$. A very simple derivation of Euler's equation

$$\zeta(3) = \frac{2\pi^2}{7}\log 2 + \frac{16}{7}\int_0^{\frac{\pi}{2}} x\log(\sin x)dx$$

and of the similar representation

$$\zeta(3) = \frac{2\pi^2}{9}\log 2 + \frac{16}{3\pi}\int_0^{\frac{\pi}{2}} x^2\log(\sin x)dx$$

was introduced. The method applies in fact to a more general case, giving us for example the formulas

$$\sum_{n=0}^{\infty}\left\{\sum_{m=0}^{n-1}\frac{1}{(2m+1)^2}\right\}\frac{1}{(2n+1)^3} = -\frac{\pi}{4}\int_0^{\frac{\pi}{2}} x^2\log(\sin x)dx + \frac{2}{3}\int_0^{\frac{\pi}{2}} x^3\log(\sin x)dx$$

and

$$\sum_{n=1}^{\infty}\left\{\sum_{m=1}^{n-1}\frac{1}{m^2}\right\}\frac{1}{n^3} = -\frac{16}{3}\int_0^{\frac{\pi}{2}} x^3\log(\sin x)dx + \frac{40}{3\pi}\int_0^{\frac{\pi}{2}} x^4\log(\sin x)dx.$$

Combinations of the above formulas may produce further interesting equations, for example

$$\sum_{n=0}^{\infty}\left\{\sum_{m=0}^{n-1}\frac{1}{(2m+1)^2}\right\}\frac{1}{(2n+1)^3} = \frac{\pi^4}{96}\log 2 - \frac{3\pi^2}{64}\zeta(3) + \frac{2}{3}\int_0^{\frac{\pi}{2}} x^3\log(\sin x)dx.$$

Finally, we wish to set the record straight by stating that we make no claim to first discovery of the correct value of $\mathrm{Li}_2(-\phi)$. We found this value on October 6, 2016, while investigating why a formula works for seven of the known dilogarithm values but fails at the eighth. Believing the result to be new, we presented it at the BG-SIAM Meeting in December 2016 (see [7]). A referee brought to our attention the fact that the value exists on the Internet (see [11]). The web-article [11] simply lists the eight dilogarithm values, and then refers the reader to Lewin's second book [6] and to the following two papers:
[A] D.H. Bailey, P.B. Borwein and S. Plouffe, On the Rapid Computation of Various Polylogarithmic Constants, *Math. Comput.* **66** (1997), 903–913,
[B] J.M. Borwein, D.M. Bradley, D.J. Broadhurst and P. Lisonek, Special Values of Multidimentional Polylogarithms, *Trans. Amer. Math. Soc.* **353** (2000), 907–941.

But in [6] the value for $\mathrm{Li}_2(-\phi)$ is wrong, and the papers [A] and [B] have no mention of any of the numbers $\mathrm{Li}_2\big(\frac{1}{\phi^2}\big), \mathrm{Li}_2\big(\frac{1}{\phi}\big), \mathrm{Li}_2\big(-\frac{1}{\phi}\big)$ or $\mathrm{Li}_2\big(-\phi\big)$. To the best of our knowledge, a discussion of the error and a proof of the correct equation $\mathrm{Li}_2\left(-\frac{\sqrt{5}+1}{2}\right) = -\frac{\pi^2}{10} - \log^2\left(\frac{\sqrt{5}+1}{2}\right)$ do not exist in the printed literature.

Acknowledgements The author expresses his gratitude to the referee who pointed out to him the fact that the value for $\mathrm{Li}_2(-\phi)$ is given on the Internet.

References

1. Borwein, J., Chamberland, M.: Integer powers of arcsin. Int. J. Math. Math. Sci., Art. ID 1981 (2007)
2. Choe, R.: An elementary proof of $\sum_{n=1}^{\infty} \frac{1}{n^2} = \frac{\pi^2}{6}$. Am. Math. Mon. **94**, 662–663 (1987)
3. Euler, L.: Exercitationes analyticae. Novi Comment. Acad. Sci. Imp. Petropol. **17**, 173–204 (1772)
4. Landen, J.: Mathematical Memoirs. London (1780)
5. Lewin, L.: Dilogarithms and Associated Functions. Macdonald, London (1958)
6. Lewin, L.: Polylogarithms and Associated Functions. Elsevier (North-Holland), New York/London/Amsterdam (1981)
7. Markov, L.: Some results involving Euler-type integrals and dilogarithm values. In: BGSIAM'16 Proceedings, pp. 54–55 (2016)
8. Maximon, L.: The dilogarithm function for complex argument. Proc. R. Soc. Lond. **459**, 2807–2819 (2003)
9. Schumayer, D., Hutchinson, D.A.W.: Physics of the Riemann hypothesis. Rev. Mod. Phys. **83**, 307–330 (2011)
10. Srivastava, H.M., Choi, J.: Series Associated with the Zeta and Related Functions. Kluwer Academic Publishers, Dordrecht/Boston/London (2001)
11. Weisstein, E.: Dilogarithm, from MathWorld—A Wolfram Web Resourse. http://mathworld.wolfram.com/Dilogarithm.html. Accessed 5 Dec 2017
12. Zagier, D.: The remarkable dilogarithm. J. Math. Phys. Sci. **22**, 131–145 (1988)

Nonlinear Evolution Equation for Propagation of Waves in an Artery with an Aneurysm: An Exact Solution Obtained by the Modified Method of Simplest Equation

Elena V. Nikolova, Ivan P. Jordanov, Zlatinka I. Dimitrova and Nikolay K. Vitanov

Abstract We study propagation of traveling waves in a blood filled elastic artery with an axially symmetric dilatation (an idealized aneurysm) in long-wave approximation.The processes in the injured artery are modelled by equations for the motion of the wall of the artery and by equation for the motion of the fluid (the blood). For the case when balance of nonlinearity, dispersion and dissipation in such a medium holds the model equations are reduced to a version of the Korteweg-deVries-Burgers equation with variable coefficients. Exact travelling-wave solution of this equation is obtained by the modified method of simplest equation where the differential equation of Riccati is used as a simplest equation. Effects of the dilatation geometry on the travelling-wave profile are studied.

1 Introduction

The theoretical investigation of pulse wave propagation in human arteries has a long history. Over the past decade the scientific efforts have been concentrated on theoretical investigations of nonlinear wave propagation through the blood in arteries with a variable radius. Clearing how local imperfections appeared in an artery can disturb the blood flow can help in predicting the nature and main features of various cardiovascular diseases, such as stenoses and aneurysms. In order to study propagation of nonlinear waves in a stenosed artery, Tay and co-authors treated the artery as a homo-

E.V. Nikolova (✉) · N.K. Vitanov
Bulgarian Academy of Sciences, Institute of Mechanics, Sofia, Bulgaria
e-mail: elena@imbm.bas.bg

N.K. Vitanov
e-mail: vitanov@imbm.bas.bg

I.P. Jordanov
University of National and World Economy, Sofia, Bulgaria
e-mail: i_jordanov@email.bg

Z.I. Dimitrova
Bulgarian Academy of Sciences, Institute of Solid State Physics, Sofia, Bulgaria
e-mail: zdim@issp.bas.bg

K. Georgiev et al. (eds.), *Advanced Computing in Industrial Mathematics*, Studies in Computational Intelligence 728,
https://doi.org/10.1007/978-3-319-65530-7_13

geneous, isotropic and thin-walled elastic tube with an axially symmetric stenosis. The blood was modeled as an incompressible inviscid fluid [1], Newtonian fluid with constant viscosity [2], and Newtonian fluid with variable viscosity [3]. Using a specific perturbation method, in a long-wave approximation the authors obtained the forced Korteweg-de Vries (KdV) equation with variable coefficients [1], forced perturbed KdV equation with variable coefficients [2], and forced Korteweg-de Vries-Burgers (KdVB) equation with variable coefficients as evolution equations [3]. The same theoretical frame was used in [4, 5] to examine nonlinear wave propagation in an artery with a variable radius. Considering the artery as a long inhomogeneous prestretched thin elastic tube with an imperfection (presented at large by an unspecified function $f(z)$), and the blood as an incompressible inviscid fluid the authors obtained again the forced KdV equation with variable coefficients. Apart from solitary propagation waves in such a system, in [5], possibility of periodic waves was discussed at appropriate initial conditions. In this text we shall focus on consideration of the blood flow through an artery with a local dilatation (an aneurysm). The aneurysm is a localized, blood-filled balloon-like bulge in the wall of a blood vessel [6]. In many cases, its rupture causes massive bleeding with associated high mortality. Motivated by investigations in [1–5], the main goal of this paper is to investigate effects of the aneurismal geometry and the blood characteristics on the propagation of nonlinear waves through an injured artery. For that purpose, we use a reductive perturbation method to obtain the nonlinear evolution equation. Exact solution of this equation is obtained by using the modified method of simplest equation. Recently, this method has been widely used to obtain general and particular solutions of economic, biological and physical models, represented by partial differential equations. The paper is organized as follows. A brief description about the derivation of equations governing the blood flow trough a dilated artery is presented in Sect. 2. In Sect. 3 we derive a basic evolution equation in long-wave approximation. A traveling wave solution of this equation is obtained in Sect. 4. Numerical simulations of the solution are presented in Sect. 5. The main conclusions based on the obtained results are summarized in Sect. 6 of the paper.

2 Mathematical Formulation of the Basic Model

It is well-known that the pulsate motion of blood causes wave propagation in arteries. In order to model the interaction of the blood with its container we shall consider two types equations which represent (i) the motion of the arterial wall and (ii) the motion of the blood. To model such a medium we shall treat the artery as a thin-walled incompressible prestretched hyperelastic tube with a localized axially symmetric dilatation. We shall assume the blood to be an incompressible viscous fluid. A brief formulation of the above-mentioned equations follows in the next two subsections.

2.1 Equation of the Wall

It is well-known, that for a healthy human, the systolic pressure is about 120 mm Hg and the diastolic pressure is 80 mm Hg. Thus, the arteries are initially subjected to a mean pressure, which is about 100 mm Hg. Moreover, the elastic arteries are initially prestretched in an axial direction. This feature minimizes its axial deformations during the pressure cycle. Experimental studies show that the longitudinal motion of arteries is very small [7], and it is due mainly to strong vascular tethering and partly to the predominantly circumferential orientation of the elastin and collagen fibers. Taking into account these observations, and following the methodology applied in [1–4], we consider the artery as a circularly cylindrical tube with radius R_0. We assume that such a tube is subjected to an initial axial stretch λ_z and a uniform (mean) inner pressure $P_0^*(Z)$ which cause relatively high circumferential and axial initial stresses. On the other hand, the pressure deviation in the course of periodic motion of heart is about ±20 mm Hg. Then the dynamical deformation due to this pressure deviation can be assumed to be smaller than the initial deformation. Therefore, the theory of small deformations superimposed on initial static deformation can be used in studying the wave propagation in such a complex medium. Under the action of such a variable pressure the position vector of a generic point on the tube can be described by

$$\mathbf{r_0} = [r_0 + f^*(z^*)]\mathbf{e_r} + z^*\mathbf{e_z},\ z^* = \lambda_z Z^* \tag{1}$$

where $\mathbf{e_r}$ and $\mathbf{e_z}$ are the unit basic vectors in the cylindrical polar coordinates, $\mathbf{r_0}$ is the deformed radius at the origin of the coordinate system, Z^* is axial coordinate before the deformation, z^* is the axial coordinate after static deformation and $f^*(z^*)$ is a function describing the dilatation geometry. We shall specify the concrete form of $f^*(z^*)$ later. Upon the initial static deformation, we shall superimpose only a dynamical radial displacement $u^*(z^*, t^*)$, neglecting the contribution of axial displacement because of the experimental observations, given above. Then, the position vector $\mathbf{r}$ of a generic point on the tube is

$$\mathbf{r} = [r_0 + f^*(z^*) + u^*]\mathbf{e_r} + z^*\mathbf{e_z} \tag{2}$$

The arc-lengths along meridional and circumferential curves respectively, are:

$$ds_z = [1 + (f^{*\prime} + \frac{\partial u^*}{\partial z^*})^2]^{1/2} dz^*,\ ds_\theta = [r_0 + f^* + u^*]d\theta \tag{3}$$

In this way, the stretch ratios in the longitudinal and circumferential directions in final configuration are

$$\lambda_1 = \lambda_z \Lambda,\ \lambda_2 = \frac{1}{R_0}(r_0 + f^* + u^*) \tag{4}$$

where

$$\Lambda = [1 + (f^{*\prime} + \frac{\partial u^*}{\partial z^*})^2]^{1/2} \tag{5}$$

The notation '$\prime$' denotes the differentiation of f^* with respect to z^*. Then, the unit tangent vector **t** along the deformed meridional curve and the unit exterior normal vector **n** to the deformed tube are

$$\mathbf{t} = \frac{(f^{*\prime} + \frac{\partial u^*}{\partial z^*})\mathbf{e_r} + \mathbf{e_z}}{\Lambda}, \quad \mathbf{n} = \frac{\mathbf{e_r} - (f^{*\prime} + \frac{\partial u^*}{\partial z^*})\mathbf{e_z}}{\Lambda} \tag{6}$$

According to the assumption made about material incompressibility the following restriction holds:

$$h^* = \frac{H}{\lambda_1 \lambda_2} \tag{7}$$

where H and h^* are the wall thicknesses before and after deformation, respectively, and λ_1 and λ_2 are the current stretch ratios in longitudinal and circumferential directions, respectively. For hyperelastic materials, the tensions in longitudinal and circumferential directions have the form:

$$T_1 = \frac{\mu^* H}{\lambda_2} \frac{\partial \Pi}{\partial \lambda_1}, \quad T_2 = \frac{\mu^* H}{\lambda_1} \frac{\partial \Pi}{\partial \lambda_2} \tag{8}$$

where $\mu^* \Pi$ is the strain energy density function of wall material as μ^* is the material shear modulus. Although the elastic properties of an injured wall section differ from those of the healthy part, here, we assume that the wall is homogeneous, i.e. μ^* is a constant through the axis z. A detailed analysis of the forces acting on an element of the artery including a free-body diagram can be found in [8, 9]. Finally, according to the second Newton's law, the equation of radial motion of a small tube element placed between the planes $z^* = const$, $z^* + dz^* = const$, $\theta = const$ and $\theta + d\theta = const$ obtains the form:

$$-\frac{\mu^*}{\lambda_z} \frac{\partial \Pi}{\partial \lambda_2} + \mu^* R_0 \frac{\partial}{\partial z^*} \left\{ \frac{(f^{*\prime} + \partial u^* / \partial z^*)}{\Lambda} \frac{\partial \Pi}{\partial \lambda_1} \right\} + \frac{P^*}{H}(r_0 + f^* + u^*)\Lambda = \rho_0 \frac{R_0}{\lambda_z} \frac{\partial^2 u^*}{\partial t^{*2}} \tag{9}$$

where t^* is the time parameter, P^* is the inner blood pressure and ρ_0 is the mass density of the tube material.

2.2 Equation of the Fluid

Experimental studies over many years demonstrated that blood behaves as an incompressible non-Newtonian fluid because it consists of a suspension of cell formed elements in a liquid well-known as blood plasma. However, in the larger arteries (with a vessel radius larger than 1 mm) it is plausible to assume that the blood has an approximately constant viscosity, because the vessel diameters are essentially larger than the individual cell diameters. Thus, in such vessels the non-Newtonian behavior becomes insignificant and the blood can be considered as a Newtonian fluid. Here, for our convenience we assume a 'hydraulic approximation' and apply an averaging procedure with respect to the cross-sectional area to the Navier-Stokes equations. Then, we obtain

$$\frac{\partial A^*}{\partial t^*} + \frac{\partial}{\partial z^*}(A^*\omega^*) = 0 \tag{10}$$

$$\frac{\partial \omega^*}{\partial t^*} + \omega^* \frac{\partial \omega^*}{\partial z^*} + \frac{1}{\rho_f}\frac{\partial P^*}{\partial z^*} = \frac{\mu_f}{\rho_f}\frac{\partial^2 \omega^*}{\partial z^{*2}} + \frac{2\mu_f}{r_f^2 \rho_f}(r\frac{\partial V_z^*}{\partial r}) \mid_{r=r_f} \tag{11}$$

where A^* denotes the inner cross-sectional area, i.e., $A^* = \pi r_f^2$ as $r_f = r_0^* + f^* + u^*$ is the final radius of the tube after deformation, ω^* is the averaged axial fluid velocity, V_z^* is the velocity component in the axial direction, ρ_f is the fluid density and μ_f is the dynamical viscosity of the fluid. The substitution of A^* in Eq. (10) leads to

$$2\frac{\partial u^*}{\partial t^*} + 2\omega^*[f^{*\prime} + \frac{\partial u^*}{\partial z^*}] + [r_0 + f^*(z^*) + u^*]\frac{\partial \omega^*}{\partial z^*} = 0 \tag{12}$$

We introduce the following non-dimensional quantities

$$t^* = (\frac{R_0}{c_0})t,\ z^* = R_0 z,\ u^* = R_0 u,\ f^* = R_0 f,\ \omega^* = c_0\omega,\ \mu_f = c_0 R_0 \rho_f \nu, \tag{13}$$

$$P^* = \rho_f c_0^2 p,\ r_0 = R_0\lambda_\theta,\ c_0^2 = \frac{\mu^* H}{\rho_f R_0},\ m = \frac{\rho_0 H}{\rho_f R_0}, V_z^* = c_0 V_z,\ r = R_0 x$$

where c_0 is the Moens-Korteweg velocity, ν is the kinematic viscosity of the fluid and λ_θ is the initial stretch ratio in a circumferential direction. We put (13) in Eqs. (12), (11) and (9), respectively. Thus the final model takes the form:

$$2\frac{\partial u}{\partial t} + 2\omega[f' + \frac{\partial u}{\partial z}] + [\lambda_\theta + f(z) + u]\frac{\partial \omega}{\partial z} = 0 \tag{14}$$

$$\frac{\partial \omega}{\partial t} + \omega\frac{\partial \omega}{\partial z} + \frac{\partial p}{\partial z} = \nu\frac{\partial^2 \omega}{\partial z^2} + \frac{2\nu}{(\lambda_\theta + f + u)^2}(\frac{\partial V_z}{\partial x}) \mid_{x=\lambda_\theta+f+u} \tag{15}$$

$$p = \frac{m}{\lambda_z(\lambda_\theta + f(z) + u)}\frac{\partial^2 u}{\partial t^2} + \frac{1}{\lambda_z(\lambda_\theta + f(z) + u)}\frac{\partial \Pi}{\partial \lambda_2}$$
$$-\frac{1}{(\lambda_\theta + f(z) + u)}\frac{\partial}{\partial z}(\frac{f' + \partial u/\partial z}{\Lambda})\frac{\partial \Pi}{\partial \lambda_1} + \nu\frac{(f' + \partial u/\partial z)\omega}{\lambda_\theta + f + u} \tag{16}$$

3 Derivation of the Evolution Equation in a Long-wave Approximation

In this section we shall use the long-wave approximation to study the propagation of waves in a fluid-solid structure system, presented by Eqs. (14)–(16). In the long-wave limit, it is assumed that the variation of radius along the axial coordinate is small compared with the wave length. As this condition is valid for large arteries, the reductive perturbation method [10] can be applied to study the asymptotic behaviour of dispersive waves in the medium. According to this method an appropriate scale transformation with a perturbation expansion of the dependent variables is introduced. The choice of coordinate transformation (known also as stretching) depends on the dispersion relationship. The dispersion relationship for such systems is derived, e.g., in [8, 9]. According to this relationship the following stretched coordinates are introduced

$$\xi = \epsilon^{1/2}(z - ct), \quad \tau = \varepsilon^{3/2} z \tag{17}$$

where ε appears in the dispersion relationship. It is a small parameter ($\varepsilon = r/l \prec 1$, where l is the characteristic wavelength) measuring the weakness of dispersion. In Eq. (17) c is the phase velocity of the harmonic wave propagation in the medium in the long-wave limit. Then, $z = \varepsilon^{-3/2}\tau$, and $f(\varepsilon^{-3/2}\tau) = \chi(\xi, \tau)$. Thus, the variables u, ω and p are functions of the variables (ξ, τ) and the small parameter ε. Taking into account the effect of dilatation, we assume f to be of order of 5/2, i.e.

$$\chi(\xi, \tau) = \varepsilon h(\tau) \tag{18}$$

In addition, taking into account the effect of viscosity, the order of viscosity is assumed to be $O(1/2)$, i.e.

$$\nu = \varepsilon^{1/2}\overline{\nu} \tag{19}$$

The last assumption ensures balance of nonlinearity, dispersion and dissipation in the system. We introduce also the following perturbation expansions of the variables u, ω and p in term of ε

$$\begin{aligned} u &= \varepsilon u_1 + \varepsilon^2 u_2 + \ldots, \ \omega = \varepsilon\omega_1 + \varepsilon^2\omega_2 + \ldots, \\ V_z &= \varepsilon V_{z1} + \varepsilon^2 V_{z2} + \ldots, \ p = p_0 + \varepsilon p_1 + \varepsilon^2 p_2 + \ldots, \end{aligned} \tag{20}$$

where $u_1 \ldots p_2$ are some unknown functions of the stretched coordinate (ξ, τ). To close the system (14)–(16) p must be presented as a function of u. Therefore we expand the other quantities in Eq. (16) in asymptotic series as follows:

$$\begin{aligned}&\lambda_1 \cong \lambda_z,\ \lambda_2 = \lambda_\theta + \varepsilon(u_1 + h) + \varepsilon^2(u_2 + (u_1 + h)^2) + \dots,\\&\frac{1}{\lambda_\theta\lambda_z}\frac{\partial\Pi}{\partial\lambda_1} = \frac{1}{\lambda_\theta\lambda_z}\frac{\partial\Pi}{\partial\lambda_z} = \gamma_0\\&\frac{1}{\lambda_\theta\lambda_z}\frac{\partial\Pi}{\partial\lambda_2} = \beta_0 + \beta_1(u_1 + h)\varepsilon + (\beta_1 u_2 + \beta_2(u_1 + h)^2)\varepsilon^2 + \dots\end{aligned} \tag{21}$$

where

$$\beta_0 = \frac{1}{\lambda_\theta\lambda_z}\frac{\partial\Pi}{\partial\lambda_\theta},\ \ \beta_1 = \frac{1}{\lambda_\theta\lambda_z}\frac{\partial^2\Pi}{\partial\lambda_\theta^2},\ \ \beta_2 = \frac{1}{2\lambda_\theta\lambda_z}\frac{\partial^3\Pi}{\partial\lambda_\theta^3} \tag{22}$$

Substituting (17)–(21) into Eqs. (14)–(16), we obtain the following differential sets:

$O(\varepsilon)$ equations

$$-2c\frac{\partial u_1}{\partial\xi} + \lambda_\theta\frac{\partial\omega_1}{\partial\xi} = 0,\ -c\frac{\partial\omega_1}{\partial\xi} + \frac{\partial p_1}{\partial\xi} = 0,\ p_1 = \gamma_1(u_1 + h) \tag{23}$$

$O(\varepsilon^2)$ equations

$$\begin{aligned}&-2c\frac{\partial u_2}{\partial\xi} + 2\omega_1\frac{\partial u_1}{\partial\xi} + \lambda_\theta\frac{\partial\omega_2}{\partial\xi} + [u_1 + h]\frac{\partial\omega_1}{\partial\xi} + \lambda_\theta\frac{\partial\omega_1}{\partial\tau} = 0\\&-c\frac{\partial\omega_2}{\partial\xi} + \omega_1\frac{\partial\omega_1}{\partial\xi} + \frac{\partial p_2}{\partial\xi} + \frac{\partial p_1}{\partial\tau} - \overline{\nu}\frac{\partial^2\omega_1}{\partial\xi^2} = 0\\&p_2 = (\frac{mc^2}{\lambda_\theta\lambda z} - \gamma_0)\frac{\partial^2 u_1}{\partial\xi^2} + \gamma_1 u_2 + \gamma_2(u_1 + h)^2\end{aligned} \tag{24}$$

From the solution of Eqs. (23), we obtain

$$u_1 = U(\xi,\tau),\ \omega_1 = \frac{2c}{\lambda_\theta}U,\ p_1 = \frac{2c^2}{\lambda_\theta}U + \gamma_1 h \tag{25}$$

where $U(\xi,\tau)$ is an unknown function whose governing equation will be obtained later. The averaged axial velocity ω_1 in Eq. (25) is determined also by a function depending on τ. However if we consider the process in infinity content this function can be removed. Comparing p_1 in Eqs. (23) and (25) leads to the following relationship $\gamma_1 = \frac{2c^2}{\lambda_\theta}$. We introduce (25) in Eqs. (24), and obtain

$$-2c\frac{\partial u_2}{\partial\xi} + \frac{4c}{\lambda_\theta}U\frac{\partial U}{\partial\xi} + \lambda_\theta\frac{\partial\omega_2}{\partial\xi} + 2c\frac{\partial U}{\partial\tau} + \frac{2c}{\lambda_\theta}(U + h)\frac{\partial U}{\partial\xi} = 0 \tag{26}$$

$$-c\frac{\partial \omega_2}{\partial \xi}+\frac{4c^2}{\lambda_\theta^2}U\frac{\partial U}{\partial \xi}+\frac{2c^2}{\lambda_\theta}\frac{\partial U}{\partial \tau}+\gamma_1 h'+\frac{\partial p_2}{\partial \xi}-\frac{4c^2}{\lambda_\theta^2}\overline{\nu}\frac{\partial^2 U}{\partial \xi^2}=0 \quad (27)$$

$$p_2=(\frac{mc^2}{\lambda_\theta \lambda z}-\gamma_0)\frac{\partial^2 U}{\partial \xi^2}+\gamma_1 u_2+\gamma_2 U^2+\gamma_2 h(\tau)U+\gamma_2 h(\tau)^2 \quad (28)$$

Replacing Eq. (28) into Eq. (27), and eliminating ω_2 between Eqs. (26) and (27), the final evolution equation takes the form:

$$\frac{\partial U}{\partial \tau}+\mu_1 U\frac{\partial U}{\partial \xi}-\mu_2\frac{\partial^2 U}{\partial \xi^2}+\mu_3\frac{\partial^3 U}{\partial \xi^3}+\mu_4(\tau)\frac{\partial U}{\partial \xi}+\mu(\tau)=0 \quad (29)$$

where

$$\mu_1=\frac{5}{2\lambda_\theta}+\frac{\gamma_2}{\gamma_1},\quad \mu_2=\frac{\overline{\nu}}{\lambda_\theta},\quad \mu_3=\frac{m}{4\lambda_z}-\frac{\gamma_0}{2\gamma_1}, \quad (30)$$
$$\mu_4(\tau)=h(\tau)(\frac{1}{2\lambda_\theta}+\frac{\gamma_2}{\gamma_1}),\quad \mu(\tau)=\frac{1}{2}h'(\tau)$$

and

$$\gamma_1=\beta_1-\frac{\beta_0}{\lambda_\theta},\quad \gamma_2=\beta_2-\frac{\beta_1}{\lambda_\theta} \quad (31)$$

Finally we have to objectify the idealized aneurysm shape. For an idealized abdominal aortic aneurysm (AAA), $h(\tau)=\delta exp(\frac{-\tau^2}{2L^2})$, where δ is the aneurysm height, i.e. $\delta=r_{max}-r_0$, and L is the aneurysm length [11]. In order to normalize these geometric quantities, we non-dimensionalize δ by the inlet radius (diameter). Then, the non-dimensional coefficient can be presented by $\delta'=DI-1$, where $DI=2r_{max}/2r_0=D_{max}/D_0$ is a geometric measure of AAA, which is known as a diameter index or a dilatation index [12]. In the same manner, the aneurysm length L is normalized by the maximum aneurysm diameter (D_{max}), i.e. $l'=L/D_{max}=1/SI$, where SI is a ratio, which is known as a sacular index of AAA [12]. For AAAs, D_{max} varies from 3 to 8.5 cm, and L varies from 5 to 10–12 cm.

4 Analytical Solution for the Nonlinear Evolution Equation: Application of the Modified Method of Simplest Equation

In this section we shall derive a travelling wave solution for the variable coefficients evolution equation, presented by Eq. (29). We shall make change of the function and the variables in the the evolution equation with variable coefficients as follows:

Let us introduce a new dependent variable such as $U(\xi,\tau) = V(\xi,\tau) - \int \mu(\tau)d\tau$. Then Eq. (29) reduces to:

$$\frac{\partial V}{\partial \tau} + \mu_1 V \frac{\partial V}{\partial \xi} - \mu_2 \frac{\partial^2 V}{\partial \xi^2} + \mu_3 \frac{\partial^3 V}{\partial \xi^3} + [\mu_4(\tau) - \mu_1 \int \mu(\tau)d\tau]\frac{\partial V}{\partial \xi} = 0. \quad (32)$$

Now, we introduce the coordinate transformation

$$\tau' = \tau,\ \xi' = \xi - \int [\mu_4(\tau) - \mu_1 \int \mu(\tau)d\tau]d\tau$$

Then, Eq. (29) is reduced to the generalized KdVB equation:

$$\frac{\partial V}{\partial \tau'} + \mu_1 V \frac{\partial V}{\partial \xi'} - \mu_2 \frac{\partial^2 V}{\partial \xi'^2} + \mu_3 \frac{\partial^3 V}{\partial \xi'^3} = 0. \quad (33)$$

Next, we shall find an analytical solution of Eq. (33) applying the modified method of simplest equation [13–16]. The short description of the modified method of simplest equation is as follows. First of all by means of an appropriate ansatz (for an example the traveling-wave ansatz) the solved of nonlinear partial differential equation for the unknown function η is reduced to a nonlinear ordinary differential equation that includes η and its derivatives with respect to the traveling wave coordinate ζ

$$\Phi\left(\eta, \eta_\zeta, \eta_{\zeta\zeta}, \dots\right) = 0 \quad (34)$$

Then the finite-series solution

$$\eta(\zeta) = \sum_{\mu=-\kappa}^{\kappa_1} a_\mu [g(\zeta)]^\mu \quad (35)$$

is substituted in (34). a_μ are coefficients and $g(\zeta)$ is solution of simpler ordinary differential equation called simplest equation. Let the result of this substitution be a polynomial of $g(\zeta)$. Equation (35) is a solution of Eq. (34) if all coefficients of the obtained polynomial of $g(\zeta)$ are equal to 0. This condition leads to a system of nonlinear algebraic equations. Each nontrivial solution of the last system leads to a solution of the studied nonlinear partial differential equation. In addition, in order to obtain the solution of Eq. (34) by the above method we have to ensure that each coefficient of the obtained polynomial of $g(\zeta)$ contains at least two terms. To do this within the scope of the modified method of the simplest equation we have to balance the highest powers of $g(\zeta)$ that are obtained from the different terms of the solved equation of kind (34). As a result of this we obtain an additional equation between some of the parameters of the equation and the solution. This equation is called a balance equation.

We introduce transformation of a traveling-wave type, i.e. $\zeta = \xi' - v^*\tau'$, where v^* is the velocity of the traveling wave. We substitute the last expression in Eq. (33)

and obtain:

$$-v^*\frac{dV}{d\zeta}+\mu_1 V\frac{dV}{d\zeta}-\mu_2\frac{d^2V}{d\zeta^2}+\mu_3\frac{d^3V}{d\zeta^3}=0. \tag{36}$$

Now we search for solution of Eq. (36) of kind $V = V(\zeta) = \sum_{r=0}^{q} a_r g^r$, where $g_\zeta = \sum_{j=0}^{m} b_j g^j$. Here a_r and b_j are parameters, and $g(\zeta)$ is a solution of some ordinary differential equation, referred to as the simplest equation. The balance equation is $q = 2m - 2$. We assume that $m = 2$, i.e. the equation of Riccati will play the role of simplest equation. Then

$$V = a_0 + a_1 g + a_2 g^2, \quad \frac{dg}{d\zeta} = b_0 + b_1 g + b_2 g^2 \tag{37}$$

The differential equation of Riccati can be written as

$$\left(\frac{dg}{d\zeta}\right)^2 = c_0 + c_1 g + c_2 g^2 + c_3 g^3 + c_4 g^4 \tag{38}$$

where

$$c_0 = b_0^2;\ c_1 = 2b_0 b_1;\ c_2 = 2b_0 b_2 + b_1^2;\ c_3 = 2b_1 b_2;\ c_4 = b_2^2 \tag{39}$$

and its solutions are given in [14]. The relationships among the coefficients of the solution and the coefficients of the model are derived by solving a system of five algebraic equations, and they are

$$a_0 = -\frac{1}{25}\frac{-3\mu_2^2 - 30\mu_2\mu_3 b_1 + 75\mu_3^2 b_1^2 + 25v\mu_3}{\mu_1\mu_3};$$

$$a_1 = -\frac{12}{5}\frac{b_2(5\mu_3 b_1 - \mu_2)}{\mu_1}; \quad a_2 = -12\frac{\mu_3 b_2^2}{\mu_1}; \quad b_0 = \frac{1}{100}\frac{25\mu_3^2 b_1^2 - \mu_2^2}{b_2\mu_3^2} \tag{40}$$

Here b_1, b_2 are free parameters. Then substituting (40) in the first equation of (37) the solution of the evolution equation with constant coefficients (Eq. (33)) is

$$V(\zeta) = -\frac{1}{25}\frac{-3\mu_2^2 - 30\mu_2\mu_3 b_1 + 75\mu_3^2 b_1^2 + 25v\mu_3}{\mu_1\mu_3} - \\ -\frac{12}{5}\frac{b_2(5\mu_3 b_1 - \mu_2)}{\mu_1}g(\zeta) - 12\frac{\mu_3 b_2^2}{\mu_1}g(\zeta)^2 \tag{41}$$

where

$$g(\zeta) = -\frac{b_1}{2b_2} - \frac{\Delta}{2b_2}\tanh\left(\frac{\Delta(\zeta+\zeta_0)}{2}\right) + \tag{42}$$
$$+\frac{\exp\left(\frac{\Delta(\zeta+\zeta_0)}{2}\right)}{2\cosh\left(\frac{\Delta(\zeta+\zeta_0)}{2}\right)\frac{b_2}{\Delta} + 2C^*\exp\left(\frac{\Delta(\zeta+\zeta_0)}{2}\right)\cosh\left(\frac{\Delta(\zeta+\zeta_0)}{2}\right)}$$

In Eq. (42) $\Delta = \sqrt{b_1^2 - 4b_0b_2} > 0$, and ξ_0 and C^* are constants of integration. The solution of the evolution equation with variable coefficients (Eq. (29)) is

$$U(\xi,\tau) = V(\zeta) - \int \mu(\tau)d\tau \tag{43}$$

where

$$\zeta = \xi - v^*\tau - \int [-\mu_1 \int \mu(\tau)d\tau + \mu_4(\tau)]d\tau \tag{44}$$

5 Numerical Findings and Discussions

It is obvious that the wave profile of the radial displacement U (Eq. (43)) depends on the material properties of the arterial wall, on the initial deformations and on the arterial geometry. In order to see their effect on the wave profile of U we need the values of coefficients $\beta_0, \beta_1, \beta_2, \gamma_0, \gamma_1, \gamma_2, \mu_1, \mu_2, \mu_3, \mu_4(\tau)$ and $\mu(\tau)$. For that purpose, the constitutive relation for tube material must be specified. Here, unlike [1–5], we assume that the arterial wall is an incompressible, anisotropic and hyperelastic material. The mechanical behaviour of such a material can be defined by the strain energy function of Fung for arteries [17]:

$$\Pi = C(e^Q - 1), \quad Q = C_1E_{QQ}^2 + C_2E_{ZZ}^2 + 2C_3E_{QQ}E_{ZZ} \tag{45}$$

where E_{QQ} and E_{ZZ} are the Green-Lagrange strains in the circumferential and axial directions, respectively, and C, C_1, C_2, C_3 are material constants. Taking into account that $E_{QQ} = 1/2(\lambda_\theta^2 - 1)$ and $E_{ZZ} = 1/2(\lambda_z^2 - 1)$, we substitute (45) in (22), (30) and (31), and obtain:

$$\begin{aligned}
\beta_0 &= \frac{1}{\lambda_z}\left(\frac{C_1}{2} + C_3(\lambda_z^2 - 1)\right)F(\lambda_\theta\lambda_z)\\
\beta_1 &= \frac{1}{\lambda_z\lambda_\theta}\left(\frac{C_1}{2} + C_3(\lambda_z^2 - 1)\right)\left(1 + \lambda_\theta^2\left(\frac{C_1}{2} + C_3(\lambda_z^2 - 1)\right)\right)F(\lambda_\theta\lambda_z)\\
\beta_2 &= \frac{1}{2\lambda_z}\left(\frac{C_1}{2} + C_3(\lambda_z^2 - 1)\right)^2\left(3 + \lambda_\theta^2\left(\frac{C_1}{2} + C_3(\lambda_z^2 - 1)\right)\right)F(\lambda_\theta\lambda_z)\\
\gamma_0 &= \frac{1}{\lambda_\theta}\left(\frac{C_2}{2} + C_3(\lambda_\theta^2 - 1)\right)F(\lambda_\theta\lambda_z),\ \gamma_1 = \frac{1}{\lambda_z}\left(\frac{C_1}{2} + C_3(\lambda_z^2 - 1)\right)^2F(\lambda_\theta\lambda_z),
\end{aligned} \tag{46}$$

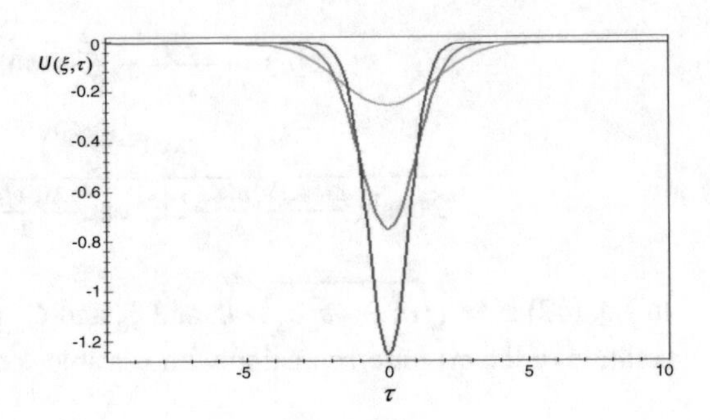

Fig. 1 Variations of the radial displacement for different values of δ' and l': for $\delta' = 0.5, l' = 1.66$ ($D_{max} = 3$ cm) (the *green line* in the figure); for $\delta' = 1.5, l' = 1$ ($D_{max} = 5$ cm) (the *red line* in the figure); for $\delta' = 2.5, l' = 0.7$ ($D_{max} = 7$ cm) (the *blue line* in the figure) ($L = 5$ cm)

$$\gamma_2 = \frac{1}{\lambda_z}(\frac{C_1}{2} + C_3(\lambda_z^2 - 1))(\frac{\lambda_\theta^2}{2}(\frac{C_1}{2} + C_3(\lambda_z^2 - 1))^2 + \frac{5}{2}(\frac{C_1}{2} + C_3(\lambda_z^2 - 1))$$
$$-\frac{1}{\lambda_\theta^2})F(\lambda_\theta \lambda_z)$$

where

$$F(\lambda_\theta \lambda_z) = C \exp(\frac{C_1}{4}(\lambda_\theta^2 - 1) + \frac{C_2}{4}(\lambda_z^2 - 1) + \frac{C_3}{2}(\lambda_\theta^2 - 1)(\lambda_z^2 - 1)) \quad (47)$$

The numerical values of material coefficients in (45) are as follows: $C = 2.5$ kPa, $C_1 = 14.5, C_2 = 7, C_3 = 0.1$. They were derived in [18] from experimental data of human aortic wall segments applying a specific inverse technique. Assuming the initial deformation $\lambda_z = 1.5, \lambda_\theta = 1.2$, we obtain the following values for the coefficients: $\beta_0 = 554.97,\ \beta_1 = 5374,\ \beta_2 = 27872.89,\ \gamma_0 = 333.36,\ \gamma_1 = 4911.52,$ $\gamma_2 = 23394.55$. Then, the numerical values of the coefficients in Eq. (29) are:

$$\mu_1 = 6.85;\ \mu_2 = 2.73.10^{-5}\text{m}^2/\text{s};\ \mu_3 = -0.017; \quad (48)$$
$$\mu_4(\tau) = 5.36\delta' \exp(-\tau^2/2l'^2),;\ \mu(\tau) = -\delta' \tau \exp(-\tau^2/2l'^2)/2l'^2.$$

We take into account that $\nu = 3.28.10^{-6}$ m^2/s when calculating μ_2. Using these numerical values, the travelling-wave solution of Eq. (29) for $\xi = 1$ is plotted in Fig. (1). In all simulations $\nu^* = 1, m = 0.1$ and $b_1 = 1, b_2 = 1$, which are defined by the symmetry condition at $\tau = 0$ and $\tau = \pm\infty$. In more detail Fig. 1 demonstrates the effect of aneurysm geometrical characteristics such as the maximal aneurysm diameter and in particular the aneurysmal length (*DI* and *SI* indexes of AAA defined in the end of Sect. 3) on the wave profile of wall displacement. Taking into account that the healthy aortic diameter is about 2 cm, various wave profiles of U are obtained for various values of the maximal aneurysm diameter D_{max} (in particular δ' or *DI*). In all these cases, a constant aneurysm length L is assumed, but l' (in particular *SI*) also varies, because D_{max} involves in this ratio. As it is seen from Fig. (1) wave elastic drop, followed by a prompt wave elastic jump is observed in presence of arterial dilatation. The graph also demonstrates that the wave amplitude increases but wave length decreases when the maximal aneurysm diameter increases. (in particular when *DI* and *SI* of AAA increase). The increasing wave amplitude of the wall displacement can lead to aneurysm rupture. Thus the obtained results are conformable with observations in the medical practice.

6 Conclusions

Modelling the injured artery as a thin-walled prestetched, anisotropic and hyperelastic tube with a local imperfection (an aneurysm), and the blood as a Newtonian fluid we have derived an evolution equation for propagation of nonlinear waves in this complex medium. Numerical values of the model parameters are determined for specific mechanical characteristics of the arterial wall and specific aneurismal geometry. We have obtained a traveling wave analytical solution of the model evolution equation. The numerical simulations of this solution demonstrate that solitary waves are observed when a local arterial dilatation appears.

Acknowledgements This work was supported by the UNWE project for scientific researchers with grant agreement No. NID NI–21/2016 and the Bulgarian National Science Fund with grant agreement No. DFNI I 02-3/12.12.2014.

References

1. Tay, K.G.: Forced Korteweg-de Vries equation in an elastic tube filled with an inviscid fluid. Int. J. Eng. Sci. **44**, 621–632 (2006)
2. Tay, K.G., Ong, C.T., Mohamad, M.N.: Forced perturbed Korteweg-de Vries equation in an elastic tube filled with a viscous fluid. Int. J. Eng. Sci. **45**, 339–349 (2007)
3. Tay, K.G., Demiray, H.: Forced Korteweg-de VriesBurgers equation in an elastic tube filled with a variable viscosity fluid. Chaos Solitons Fractals **38**, 1134–1145 (2008)
4. Demiray, H.: Non-linear waves in a fluid-filled inhomogeneous elastic tube with variable radius. Int. J. Non-Linear Mech. **43**, 241–245 (2008)
5. Dimitrova, Z.I.: Numerical investigation of nonlinear waves connected to blood flow in an elastic tube of variable radius. J. Theor. Appl. Mech. **45**, 79–92 (2015)
6. Aneurysm, From Wikipedia, the free encyclopedia. https://en.wikipedia.org/wiki/Aneurysm
7. Patel, P.J., Greenfield, J.C., Fry, D.L.: In vivo pressure length radius relationship in certain blood vessels in man and dog. In: Attinger, E.O. (ed.) Pulsatile Blood Flow, p. 277. McGraw-Hill, New York (1964)
8. Demiray, H.: Wave propagation though a viscous fluid contained in a prestressed thin elastic tube. Inf. J. Eng. Sci. **30**, 1607–1620 (1992)
9. Demiray, H.: Waves in fluid-filled elastic tubes with a stenosis: variable coefficients KdV equations. J. Comput. Appl. Math. **202**, 328–338 (2005)
10. Jeffrey, A., Kawahara, T.: Asymptotic methods in nonlinear wave theory. Pitman, Boston (1981)

11. Gopalakrishnan, S.S., Benot, P., Biesheuvel, A.: Dynamics of pulsatile flow through model abdominal aortic aneurysms. J. Fluid Mech. **758**, 150–179 (2014)
12. Raut, S.S., Chandra, S., Shum, J., Finol, E.A.: The role of geometric and biomechanical factors in abdominal aortic aneurysm rupture risk assessment. Ann. Biomed. Eng. **41**, 1459–1477 (2013)
13. Vitanov, N.K., Dimitrova, Z.I., Kantz, H.: Modified method of simplest equation and its application to nonlinear PDEs. Appl. Math. Comput. **216**, 2587–2595 (2010)
14. Vitanov, N.K.: Modified method of simplest equation: powerful tool for obtaining exact and approximate traveling-wave solutions of nonlinear PDEs. Commun. Nonlinear Sci. Numer. Simul. **16**, 1176–1185 (2011)
15. Vitanov, N.K.: On modified method of simplest equation for obtaining exact and approximate solutions of nonliear PDEs: the role of simplest equation. Commun. Nonlinear Sci. Numer. Simul. **16**, 4215–4231 (2011)
16. Vitanov, N.K., Dimitrova, Z.I., Vitanov, K.N.: Modified method of simplest equation for obtaining exact analytical solutions of nonlinear partial differential equations: further development of the methodology with applications. Appl. Math. Comput. **269**, 363–378 (2015)
17. Fung, Y.: Biomechanics: Mechanical Properties of Living Tissues. Springer, New York (1993)
18. Avril, S., Badel, P., Duprey, A.: Anisotropic and hyperelastic identification of in vitro human arteries from full-field optical measurements. J. Biomech. **43**, 2978–2985 (2010)

Modelling of Light Mg and Al Based Alloys as “in situ” Composites

Ludmila Parashkevova and Pedro Egizabal

Abstract The present paper is aimed to further elucidate the microstructure properties relationship of light alloys containing additional hard particles. The materials studded are magnesium alloys from the system AZ (Mg–Al–Mn–Zn) and mechanically alloyed aluminum reinforced with carbide and oxide particles. Strengthening and hardening phenomena in Metal Matrix Multiphase heterogeneous Materials (MMMM) are considered in this study from the view point of mechanics of nano- and micro-composites. A semi-analytical approach is adopted taking into account the manufacturing processing and microstructure features. Multilevel homogenization procedure is performed, accounting for size effects. In the model applied the metal matrix is considered as an elastic–plastic micropolar media and the hard phases (precipitations Mg17Al12, TiC, Al4C3, Al2O3) are treated as conventional elastic Cauchy materials. Experimentally observed dependence of the characteristic matrix length on the volume fraction of the hardening phases is modeled and numerically simulated in the case of ball-milled Al alloyed with Al4C3 and Al2O3. For AZ alloys the impact of intermetallic phase Mg17Al12 is discussed in the frame of presented composite model and the strengthening effect of the addition of small amount of TiC is estimated.

1 Introduction

In recent years, the development of enhanced technologies for preparing metal matrix composites with unique properties shows a significant increase. The resulting new materials exhibit high mechanical properties and remarkable thermal and structural stability in operation conditions, [14]. This paper is aimed to elucidate some problems of metal material strengthening caused by an additional hard phase appear-

L. Parashkevova (✉)
Institute of Mechanics, Bulgarian Academy of Sciences, Sofia, Bulgaria
e-mail: lusy@imbm.bas.bg

P. Egizabal
Foundry and Steelmaking Department, TECNALIA, Donostia - San Sebastián, Spain
e-mail: pedro.egizabal@tecnalia.com

K. Georgiev et al. (eds.), *Advanced Computing in Industrial Mathematics*, Studies in Computational Intelligence 728,
https://doi.org/10.1007/978-3-319-65530-7_14

ance. Two of the physical—chemical processes connected to hardening in metals and alloys, are:

- release of precipitations from the parent melt during solidification;
- synthesis of metal carbides/oxides/silicides during mechanical alloying.

We apply herein a non-standard approach to estimate the impact of the mentioned phenomena regarding the metals and alloys as natural "in situ" composites. These heterogeneous materials are considered as multiphase metal matrix composites. The basic elastic–plastic properties of such materials are predicted and valuated by means of a modified variant of homogenization theory coupled with phenomenological relations describing micro- macro-structure.

The approach is applied for analysis of two types of light alloys. The first type are magnesium (Mg) based AZ alloys with aluminum (Al) content (1–10 wt%). The main representative of the AZ family is the AZ91 alloy. It contains a higher percentage of aluminium (around 9 wt% Al) than the AM-alloys and about 0.7 wt% zinc. The major advantage, which keeps "awake" the interest in Mg alloys is their lower density directly connected to a weight saving of about 40% compared to steel and cast iron and 20% compared to aluminum for the same component performance, [18].

The second type of light alloys are mechanically alloyed pure Al containing carbides $Al4C3$ and oxcides $Al2O3$ (0–10 vol. %). Al-based alloys have a high strength/weight ratio, good formability, excellent combination of castability and mechanical properties which together with an excellent corrosion resistance make them very appropriate for a large variety of applications, [18]. Various possible hardening—strengthening effects in both types of light alloys are presented and discussed.

2 Microstructure and Experimental Observations

For Mg cast alloys of the system AZ α Mg matrix is the predominate phase and the main precipitate phase is Mg17Al12 (γ phase). The phase γ is thermally rather unstable but is much harder than the matrix Mg phase, [7], Young's moduli ratio ranges the interval (1.5–1.85). The bulk moduli of both phases, however, are very close. This feature of two-phase Mg–Al alloy considered as a "in situ" composite leads to almost equal volume changes under elastic deformation on microstructural level and slight discrepancy and low crack potential on macro level. It was revealed that discontinuous and continuous precipitates can occur independently, simultaneously or competitively, dependently on the ageing (cooling) regime, [3]. Discontinuous precipitates mainly locate at the grain boundaries, the continuous precipitate phase appears in various growth directions on the interface and the inner of grains as well. It is proved that continuous precipitation tends to be favoured at high temperatures (i.e. close to the solvus curve) whereas at low temperatures of ageing, discontinuous precipitation prevails. Spherical shape of γ precipitates can be obtained due to correlate precipitation and magnesium matrix deformation processes, [3].

Each real solidification path depends on temperature decreasing rate, on the mass diffusivities of the solutes in liquid and back into the primary solid phase as well. It lies between two limiting cases. The first one is a solidification path at global equilibrium conditions, represented by the levers rule. The second one corresponds to the process with no diffusion in the solid phase and is described by Scheil-Gulliver model, [10]. For Mg–Al casting alloys in solid state the contribution of intermetallic phase to strength not only depends on its fraction but also on its morphology. The strengthening effect is higher when the intermetallic phase is interconnected rather than being lamellar and discontinuous, [7]. The supersaturation of solution due to high diffusion rate of Al in Mg change the balance between solid phases and could account for another strengthening of the AZ alloys.

Mechanical Alloying (MA) is a solid-state powder manufacturing technique, developed to combine oxide strengthening, precipitation and dispersion hardening. Starting from a set of specially blended compounds, the powder mixture is subjected to severe mechanical collision treatment like ball-milling or hot pressing followed by consolidation processing up to bulk homogeneous state. In comparison to other technologies mechanical alloying is a relatively simple process leading to considerable advantages such as fine final microstructure and high volume fraction of reinforcement phase introduced into the composite.

The methods of MA are very effective for obtaining ultra fine microstructure up to nano size leading to higher strengthening, better ductility, fracture toughness and high temperature resistance. After MA the reinforcement phases are spread both inside the grains and along the grain boundaries in a manner similar to that of continuous precipitating in aged cast alloys.

As other nanocomposites MA composites demonstrate significant size sensitivity, depending on the processing technology and on the working conditions. Some recent investigations on MA report that a clear relationship exists between hardening phase parameters (size and volume fraction) and microstructure characteristics of the matrix, [4, 9]. In particular [4] shows that the microstructure evolution of the aluminum matrix is manifested by changes of the mean crystallite size and dislocation density depending on the volume fraction of the strengthening particles.

3 Modelling

Different theoretical models accounting for hardening of particle enforced alloys have been proposed during last decades [12, 14]. They are based on thermodynamic, kinetic and dislocation mechanisms. The particles of the dispersed or precipitated second phase, integrated into the matrix in different physical-chemical ways (coherent or non-coherent) are considered as obstacles for dislocation transportation. This classical approach presumes that the overall yield strength of the alloy is described by an additive function:

$$\sigma_{YS} = \sigma_m + \sigma_{SS} + \sigma_{GS} + \sigma_d + \sigma_p\ , \tag{1}$$

where σ_m, σ_{SS}, σ_{GS}, σ_d, σ_p are the corresponding contributions of the matrix itself, of the solid solution, of the grain size hardening, of the dislocation hardening and of the hardening due to precipitations or inclusions. Relation (1) can be presented in the form $\sigma_{YS} = \sigma_0 + M\Delta\tau$ where $\sigma_0 = \sigma_m + \sigma_{SS} + \sigma_{GS} + \sigma_d$, M is Taylors factor, $M = 3$ is reference value for metals; $\Delta\tau$ is the increase of the shear stress of the material. By means of the term σ_{GS} these expressions include also Hall-Petch type relations in which hardening effect depends on grain size D_g:

$$\sigma_{YS} \sim \sigma_m + A(D_g)^{-1/2} \quad A = const\ . \tag{2}$$

To extend the classical Cauchy continuum mechanics theory to a higher-order medium, it is assumed that any material point is endowed with an internal microstructure, [8]. The Cosserat continuum belongs to the larger class of generalized continua which introduce intrinsic length scales into continuum mechanics via higher order gradients and additional degrees of freedom of fully non local constitutive equations, [11, 15]. Three displacements are employed in the usual way to characterize the macroscopic motion of the material point, and three additional microrotation angles are introduced to describe the rotation of the microstructure within the material point. The recent renewal of Cosserat mechanics is due to the dramatic increase of computational capabilities, to the development of local strain field measurement methods and to the enormous interest in size effects in modern micro- and nano-structure materials. The dependence of the effective properties of metal matrix composites on the size of the particles or fibers can be accounted for by treating the matrix as a Cosserat or a generalized medium.

A mechanical model, appropriate for nano–micro composites should account for two kinds of size effects: (a) size effects due to particles themselves (b) size effects due to particles matrix mutual influence.

On micro level the hardening phases in a composite alloy are assumed uniformly dispersed spheres (equivalent inclusions) with diameter D_i and given mechanical properties. They can be grouped in a finite number n of sets, according to their size and mechanical properties. The volume fraction of each set in the RVE (Representative Volume Element) is C_i. The total volume fraction of the hardening phases (equivalent inclusions) is $C_{sum} = \sum_{i=1}^{n} C_i$. The material of the inclusions is Cauchy-type elastic isotropic one with mechanical characteristics: Young's modulus E_i, Poisson's ratio ν_i, shear modulus G_i and bulk modulus K_i. The matrix of the multiphase composite is considered as centro-symmetric micropolar elastic–plastic work-hardening continuum. The stress and strain measures are the stress tensor $\sigma_{ij} = \sigma_{(ij)} + \sigma_{\langle ij \rangle}$, the couple stress tensor $m_{ij} = m_{(ij)} + m_{\langle ij \rangle}$, the strain tensor $\varepsilon_{ij} = \varepsilon_{(ij)} + \varepsilon_{\langle ij \rangle}$ and the curvature tensor $k_{ij} = k_{(ij)} + k_{\langle ij \rangle}$. Symbols $(\ldots)$ and $\langle \ldots \rangle$ in the subscript denote the symmetric and anti-symmetric parts of a tensor respectively.

The elastic behaviour of the matrix on micro level is described with the well known relations which can be found in [15, 17]. According to micropolar elasticity the model parameters are: α, β, γ and κ are the Cosserat material constants. K_0 and N are Cauchy and Cosserat bulk moduli respectively, E_0 is Young's modulus and ν_0 is Poisson's ratio. G_0 is Cauchy shear modulus. The matrix characteristic length parameter l_m is defined in accordance with the structure of the constitutive relations as: $l_m^2 = \alpha/\lambda_0 = \beta/G_0 = \gamma/\kappa$. Usually the length parameter l_m is associated with grain size or crystallite size in metals and alloys. Everywhere in the paper $(\ldots)'$ means deviator. We assume the following equivalent stress $\overline{\sigma}_e$ [6] for describing the transition from elastic to plastic state

$$\overline{\sigma}_e^2 = \frac{3}{2}\sigma'_{(ij)}\sigma'_{(ij)} + \frac{3}{2l_m^2}m'_{(ij)}m'_{(ij)} + \frac{3}{2l_m^2}m_{\langle ij\rangle}m_{\langle ij\rangle} \tag{3}$$

and corresponding yield condition for the matrix on micro level:

$$\overline{\sigma}_e = \sigma_{p0}\left(\overline{\varepsilon}_p\right) = \sigma_{p0}^0 + h_0\overline{\varepsilon}_p^{\ m}, \tag{4}$$

where σ_{p0} is the yield stress of the matrix, σ_{p0}^0 is its initial value, h_0 and m are hardening parameters. $\overline{\varepsilon}_p$ is the equivalent plastic strain on microlevel. According to the decomposition approach [19] the multiphase RVE, consisting of matrix and n_f phases, is equivalent to a RVE, consisting of n pseudograins, $n \geq n_f$. Each pseudograin is a two-phase composite, built of part of the matrix and all inclusions with a particular size and elastic properties. The volume of matrix is distributed among pseudo grains proportionally to the volume of each hardening phase. As a result the volume fractions of inclusions are equal to C_{sum} for all pseudo grains. The volume fraction of the pseudograin i with respect to RVE is $\tilde{C}_i = C_i/C_{sum}$.

Two–steps homogenization procedure is performed as a key to proceed from micro to macro level. On the first step the material of each pseudo grain should be homogenized as a two-phase composite following a proper scheme which depends on the total volume fraction of the inclusions. On the second step the RVE's agglomerate of the already homogeneous Cauchy—type pseudo-grains has to be subjected to the final homogenization. In the case of low volume fraction, not exceeding 20–30%, the hypothesis of dilute inclusions is valid for each pseudo-grain and updated size sensitive Mori–Tanaka homogenization could be applied. For the RVE of the metal composites considered the moduli of i-th pseudo-grain are obtained by means of equations:

$$K_{ci} = K_0\left[1 + \frac{C_{sum}\left(K_i - K_0\right)}{C_0 a_0\left(K_i - K_0\right) + K_0}\right],\ G_{ci} = G_0\left[1 + \frac{C_{sum}\left(G_i - G_0\right)}{C_0 b_{0i}\left(G_i - G_0\right) + G_0}\right], \tag{5}$$

where $a_0 = a_0(G_0, K_0)$, $b_{0i} = b_{0i}(G_0, K_0, D_i, l_m)$, [6]. The size sensitivity of the model depends on the dimensionless parameter D_i/l_m through the average Eshelby tensor component b_{0i}, [17]. It is important to emphasize that when $0 \leq D_i/l_m \leq 1$ the strengthening effects are the most pronounced, but limited.

For the second homogenization we chose the self-consistent theory for polycrystalls insofar as all pseudo-grains must be treated in a similar way and only a symmetric homogenization scheme is suitable. The overall moduli of composite are calculated numerically as a solution of the non-linear equation system:

$$\sum_{i=1}^{n}\frac{\tilde{C}_i}{1-\frac{3\overline{K}}{3\overline{K}+4\overline{G}}\left(1-\frac{K_{ci}}{\overline{K}}\right)}-1=0,\quad \sum_{i=1}^{n}\frac{\tilde{C}_i}{1-\frac{2\left(3\overline{K}+6\overline{G}\right)}{5\left(3\overline{K}+4\overline{G}\right)}\left(1-\frac{G_{ci}}{\overline{G}}\right)}-1=0\,. \tag{6}$$

We assume that the plastic stage of the composite is reached if the yield condition for the matrix material is satisfied in a averaged manner, as far as the inclusions are considered pure elastic and do not undergo plastic deformation. The following yield condition on macro level is proposed:

$$\langle\overline{\sigma}_e{}^2\rangle_0=\sigma_{p0}^2\left(\langle\overline{\varepsilon}_p\rangle_0\right)\,. \tag{7}$$

The variation technique of Hu et al. [6], described in [16], is used to evaluate the averaged equivalent stress $\langle\overline{\sigma}_e{}^2\rangle_0$. The concept of the secant moduli method is applied in the inelastic state. According to it at each deformation step the real nonlinear material is compared with an elastic one, having diminishing elastic properties.

$$E_0^s=\left(\frac{1}{E_0}+\frac{\overline{\varepsilon}_p}{\sigma_{p0}}\right)^{-1},\quad \nu_0^s=\frac{1}{2}-\left(\frac{1}{2}-\nu_0\right)\frac{E_0^s}{E_0}, \tag{8}$$

Keeping in mind (7) the yield condition of the overall composite material is rewritten in the form:

$$F\equiv\frac{3}{2}\Sigma'_{ij}\Sigma'_{ij}+\frac{\overline{A}_c{}^2}{9\overline{B}_c{}^2}\Sigma_{kk}^2-\sigma_{pc}^2=0 \tag{9}$$

$$\sigma_{pc}=\overline{A}_c\sigma_{p0}\left(\langle\overline{\varepsilon}_p\rangle_0\right)=\overline{A}_c\left[\sigma_{p0}^0+h_0\left(\overline{E}_p/C_0\right)^m\right] \tag{10}$$

where the yield limit of the composite σ_{pc} depends on the relation between equivalent plastic strain measures on micro and macro levels. Equation (10) is valid until the inelastic deformations of harder phases could by neglected. The coefficients $\overline{A}_c$ and $\overline{B}_c$ are given with:

$$\begin{aligned}\frac{1}{\overline{A}_c{}^2}&=\frac{1}{C_0}\left(\frac{G_0{}^2}{\overline{G}^2}\frac{\partial\overline{G}}{\partial G_0}+\frac{1}{l_m^2}\frac{\beta^2}{\overline{G}^2}\frac{\partial\overline{G}}{\partial\beta}+\frac{1}{l_m^2}\frac{\gamma^2}{\overline{G}^2}\frac{\partial\overline{G}}{\partial\gamma}\right),\\ \frac{1}{9\overline{B}_c{}^2}&=\frac{1}{3C_0}\left(\frac{G_0{}^2}{\overline{K}^2}\frac{\partial\overline{K}}{\partial G_0}+\frac{1}{l_m^2}\frac{\beta^2}{\overline{K}^2}\frac{\partial\overline{K}}{\partial\beta}+\frac{1}{l_m^2}\frac{\gamma^2}{\overline{K}^2}\frac{\partial\overline{K}}{\partial\gamma}\right).\end{aligned} \tag{11}$$

The derivatives appearing in (11) can be found in [17] in details. At transition from elastic to plastic state the elastic moduli of the matrix in (5), (6), (9) and (11) should be replaced by the matrix secant ones, described by (8).

If the bulk moduli of the composite constituents are equal ($K_0 = K_i,\ i = 1, \ldots, n$) than $1/\overline{B}_c^{\,2} = 0$ and the overall material obeys von Mises yield condition, see (9).

4 Numerical Simulations and Results

In classical micromechanics of composites the characteristic parameters of the matrix and additional phases are considered constant and independent of each other. In this study we test the presented methodology at different conditions. Using the microstructural observations and phenomena mentioned in Sect. 2 special relationships among model parameters are suggested and their influence on mechanical properties on macro level is investigated.

4.1 Properties of Mg Alloys from AZ System

As a first task we are modelling Mg alloy of type AZ with matrix αMg phase and continuously precipitated γ Mg17Al12 phase as a two-phase composite. According to mass conservation low one can derive the following relation among fractions of the AZ alloy in solid state, regarding Al redistribution:

$$C_{Al}^{wt\alpha} = C_{Al}^{wt\gamma} - \frac{C_{Al}^{wt\gamma} - C_{Al}^{wtAZ}}{\left[1 - \left(1 - \frac{\rho^{\alpha Mg}}{\rho^{AZ}}\right) C_{\gamma}^{volAZ}\right]\left(1 - C_{\gamma}^{volAZ}\right)}, \tag{12}$$

where $\rho^{(.)}$ stays for density, $C_{(..)}^{wt\,(.)}$, $C_{(..)}^{vol(.)}$ stays for mass (volume) fraction of the component (..) in (.). In (12) $\rho^{\alpha Mg}$, C_{γ}^{volAZ} are unknown, because nominal amount of Al could be distributed between both phases of AZ alloy in different ways. To overcome this uncertainty we suggest the following additional relationships:

$$\rho^{\alpha Mg} = \rho^{AZ}\left(1 - C_{\gamma}^{volAZ}\right) + \rho^{Mg} C_{\gamma}^{volAZ} \tag{13}$$

$$C_{\gamma}^{volAZ} \cong C_{\gamma\,max}^{volAZ}(Al)\,\frac{T_{solvus} - T}{T_{solidus} - T} = C_{\gamma\,max}^{volAZ}\left(1 - \frac{T_{solidus} - T_{solvus}}{T_{solidus} - T}\right) \tag{14}$$

Relation (14) expresses the assumption that the higher is the content of Al in AZ alloy, the higher is the temperature of decomposition of solid solution, the lower is the potential of α phase to keep the saturated Al. All this contributes to possible increase of precipitation phase. The Mg—rich part of Mg–Al phase equilibrium diagram is used herein with $T = T_{room}$ as a reference temperature. It is clear that Eq. (14) should be modified if any data about the γ phase volume fraction is available from a particular solidification path or additional thermal treatments. The maximum volume fraction of γ phase can be distinguished if there is no trace of Al rest in the α phase, i.e. from (12) one gets:

$$0 = C_{Al}^{wt\gamma} - \frac{C_{Al}^{wt\gamma} - C_{Al}^{wtAZ}}{\left[1 - \left(1 - \frac{\rho^{Mg}}{\rho^{AZ}}\right) C_{\gamma\, max}^{volAZ}\right]\left(1 - C_{\gamma\, max}^{volAZ}\right)} \tag{15}$$

$$C_{\gamma\, max}^{volAZ} = \frac{1 + a_1 - \sqrt{(1 + a_1)^2 - 4a_1 a_2}}{2a_1}, \; a_1 = 1 - \rho^{Mg}/\rho^{AZ}, \; a_2 = C_{Al}^{wtAZ}/C_{Al}^{wt\gamma} \tag{16}$$

Analyzing experimental data for tensile behavior of different AZ alloys presented in [1, 5] it is supposed that the yield strength of α phase depends on Al content through the relation:

$$\sigma_{p0} = \sigma_{02}^{\alpha Mg} = S_0 \left(1 + S\, C_{Al}^{wt\alpha}\right), \tag{17}$$

where $S_0 = 56$ MPa, $S = 17$. The elastic moduli data from [7] are incorporated into the model and we calculate the input parameters needed for two-steps homogenization varying the Al content from zero to 12.5 wt%, see Table 1. The predictions from Eqs. (12) and (14) are illustrated on Fig. 1 and are used to estimate the influence of Al redistribution on the initial yielding of matrix phase and the impact of harder γ phase on yield stress of AZ alloys, which contain only one precipitated phase, see Fig. 2.

The next example we consider is a composite consisting of AZ91 D alloy die-cast with addition of small amount of TiC particles. Data presented in Table 2 for AZ91 alloy (as matrix material) and for TiC (as a second hard phase) are experimentally obtained. On Fig. 3 the numerically simulated elastic–plastic behavior of the composite on macro structural level is shown. We studded two variants accounting for TiC particles size: one is by average diameter, given in the Table 2, and other one introducing (by means of size distribution density function) full range of measured diameters of TiC. As far as for all observed sizes the size sensitivity parameter $D_i/l_m \ll 1$, the hardening effect in both cases is numerically identical (Fig. 3).

Table 1 Input parameters for simulation of AZ alloy as a composite

Material	E (GPa)	ν	κ (GPa)	σ_{p0} (MPa)	l_m (μm)	D_γ (μm)
αMg	44	0.33	16.5	Equation (17)	100	–
$Mg17Al12$	71.94	0.23	–	–	–	10

4.2 Properties of Mechanically Alloyed Al Hardened by Al4C3 and Al2O3

The material under consideration is powder mixture of Al and carbon C subjected to ball-milling followed by hot extrusion. In modelling the end bulk material is regarded as a three-phase composite consisting of micropolar elastic–plastic Al matrix and two Cauchy elastic phases Al4C3 and Al2O3. It is supposed that during manufacturing the matrix has been forced to accommodate with the presence of inclusions of different nature by forming fine crystallites among particles. Roughly speaking Al matrix "builds" its crystallites in accordance to the interparticle distances. Let remember that Cosserat length parameter is related to matrix microstructure so the following relation is adopted:

$$l_m = \langle L \rangle, \tag{18}$$

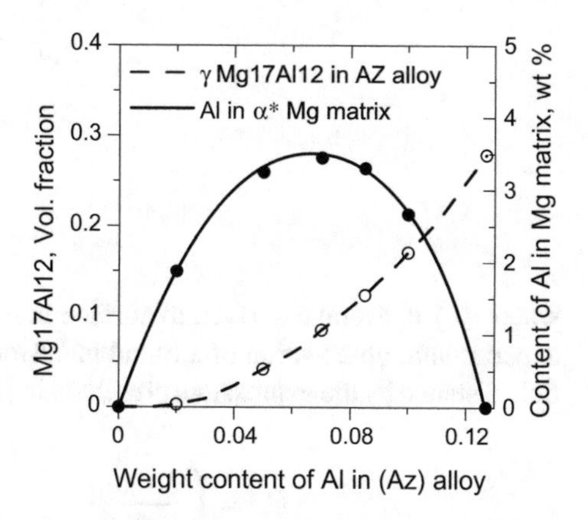

Fig. 1 Influence of Al content in AZ alloy on its compounds

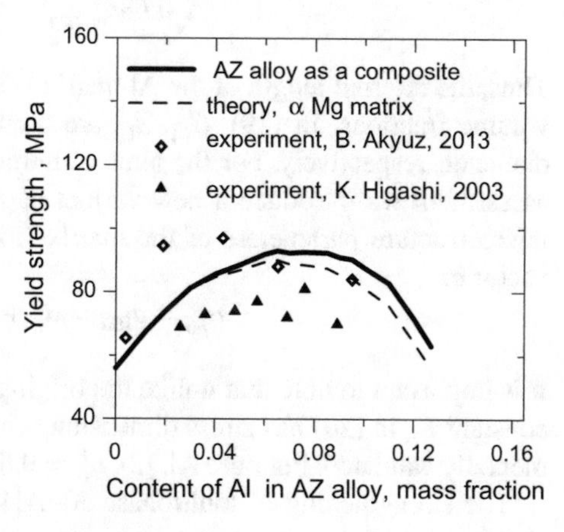

Fig. 2 Tensile yield limit of Mg based AZ alloys in case of nominal Al content distribution corresponding to Fig. 1

Table 2 Input parameters for (AZ91 + TiC) composite

Material	E (GPa)	ν	κ (GPa)	σ^0_{p0} (MPa)	h_0 (MPa)	m power	l_m (μm) (D_{TiC})	Fraction vol. (%)
AZ91	45	0.35	16.67	99	450	0.5	22	99.43
TiC	440	0.189	–	–	–	–	0.93	0.57

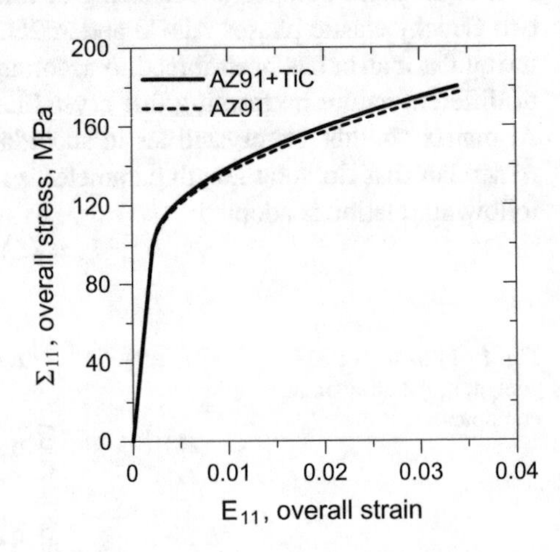

Fig. 3 Overall elastic–plastic behaviour of AZ91 alloy composite with and without 0.0057% TiC particles

where $\langle L \rangle$ is average surface to surface interparticle distance. In conformity with experimental observation of a round inclusions shape we approximate the interparticle distance by the expression, presented in [13] for the case of spherical inclusions:

$$\langle L \rangle = \left(\frac{C_{Al4C3}}{0.68 d^3_{Al4C3}} + \frac{C_{Al2O3}}{0.68 d^3_{Al2O3}} \right)^{-1/3} \tag{19}$$

Thus, the internal length of the Al matrix is conjugated with other phases sizes and volume fractions. In (19) $C_{(.)}$, $d_{(.)}$ are corresponding phase volume fraction and diameter, respectively. For the aims of homogenization and mechanical properties assessment we introduce a new variant of Hall–Petch equation (2) involving two microstructure parameters of the matrix Cosserat internal length l_m and Burger's vector b:

$$\sigma^0_{p0} = \sigma_{0Al} = \sigma^p_0 + K_b \sqrt{b} / \sqrt{l_m} \tag{20}$$

It is important to note that unlike the original form of Hall–Petch, equation (2), the constant K_b in (20) has stress dimension. This could be important if (20) is experimentally validated. For pure Al, [2], $\sigma^p_0 = 9.8$ MPa, $K_b = 3662$ MPa, $b = 0.2865$ nm.

The strengthening of multiphase Al–Al4C3–Al2O3 composite is modeled and simulated applying the processing and compounds parameters, given in [4] and Table 3. On Fig. 4 is demonstrated that the relations (18) and (19) adopted in the

Table 3 Properties of Al based composite

Properties	Al matrix	Inclusions	
		Al4C3	*Al2O3*
Young's modulus (GPa)	71	411	393
Poisson's ratio	0.34	0.24	0.22
Vol. fraction (%)	89–98	1–10	1
κ (GPa)	26.0	**	–
l_m; part. diameter (nm)	$l_m = \langle L \rangle$	40	500

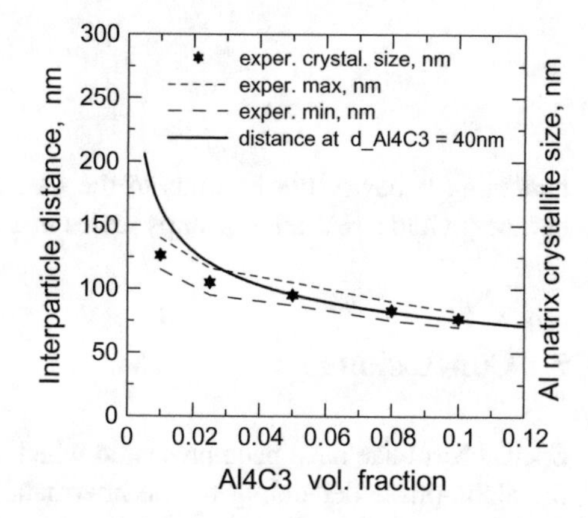

Fig. 4 Dependence of crystallites size in Al matrix and average interparticle distance on Al carbide inclusions volume fraction

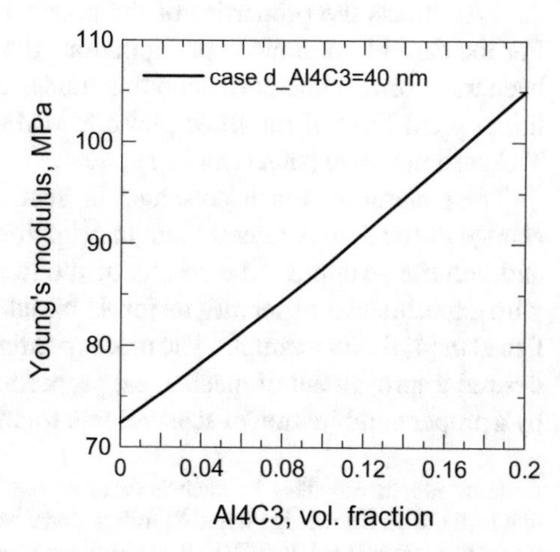

Fig. 5 Dependence of overall elastic modulus of Al composite alloy on Al carbide inclusions volume fraction, Al4C3 particle size 40 nm

present model, rather well describe the experimental observations of the crystallite size dependence on hardening phase volume fraction. Using the relation $Hv\,[kG/mm_2] = 2.9 \times 9.8\,\sigma_{pc}$ [MPa], verified for aluminum alloys, and hardening measurement data, presented in [4] one could see that model predicted elastic and

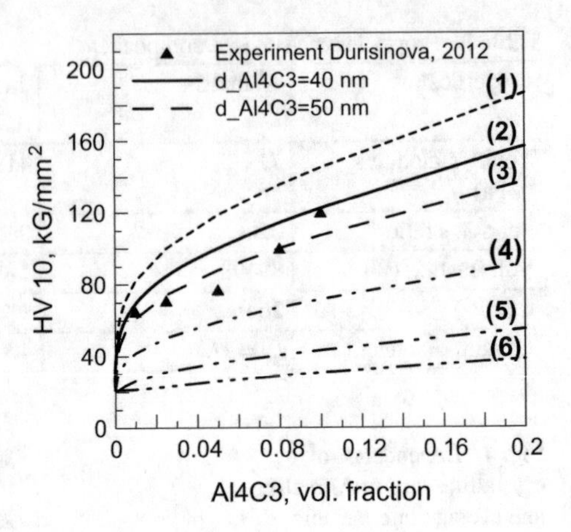

Fig. 6 Hardening of end composite (Al + Al4C3 + Al2O3). Influence of Al4C3 particle size: *curve (1)* 30 nm, *curve (2)* 40 nm, *curve (3)* 50 nm, *curve (4)* 100 nm, *curve (5)* 250 nm, *curve (6)* 500 nm

hardening behavior corresponds to the real one, (Figs. 5 and 6), even though the hardness yield stress relationships varies in a wide range.

5 Conclusions

Special formulae have been presented which estimate the amount of the Al kept in the alpha-phase depending on the intermetallic phase precipitated. Such a "residual" Al affects the properties of the matrix phase and of the AZ alloy as a whole. For the case of continuous precipitations the chosen homogenization approach has been transformed into corresponding numerical routines. The elastic moduli and the initial yield limit of the three-phase MMMM based on AZ91D alloy modified by TiC particles have been obtained.

The general approach presented in Sect. 3 is modified to take into account the change of the matrix internal length properties depending on the precipitations size and volume fractions. The results of the numerical simulations are in accordance with experimental observations for Al based mechanically alloyed composites published in [4, 9], for example. The model predictions outline some new areas where the desired improvement of mechanical properties of the end materials can be achieved by a proper combination of size, volume fraction and properties of hardening phases.

Acknowledgements This research is carrying out in the frame of KMM-VIN—European Virtual Institute on Knowledge-based Multifunctional Materials AISBL. Partial support from BG FSI through the project H 07/37/2016 is gratefully acknowledged.

References

1. Akyuz, B.: Influence of Al content on machinability of AZ series Mg alloys. Trans. Nonferrous Met. Soc. China **23**, 2243–2249 (2013)
2. Armstrong, R.W.: Yield, flow and fracture of polycrystalls. In: Baker, T.N. (ed.) Applied Science Pub., pp. 1–31. London (1983)
3. Braszczynska-Malik, K.: Precipitates of Gamma-Mg17Al12 phase in Mg–Al Alloys. In: Czerwinski, F. (ed.) Magnesium Alloys—Design, Processing and Properties, Chapter 5, pp. 95–112. InTech (2011)
4. Durisinova, K., Durisin, J., Orolinova, M., Durisin, M.: Effect of particle additions on microstructure evolution of aluminum matrix composite. J Alloy Compd. **525**, 137–142 (2012)
5. Higashi, K., Hirai, K.: Method of manufacturing magnesium alloy products, Patent US 20030173005 A1. https://www.google.ch/patents/US20030173005 (2003)
6. Hu, G., Liu, X., Lu, T.J.: A variational method for non-linear micropolar composites. J. Alloy Compd. **37**, 407–425 (2005)
7. Huang, Z.W., Zhao, Y.H., Hou, H., Han, P.D.: Electronic structural, elastic properties and thermodynamics of Mg17Al12, Mg2Si and Al2Y phases from first-principles calculations. Physica B **407**, 1075–1081 (2012)
8. Kanoute, P., Boso, D.P., Chaboche, J.L., Schrefler, B.A.: Multiscale methods for composites. Arch. Comput. Method E **16**, 31–75 (2009)
9. Kulkov, S.N., Vorozhtsov, S.A.: Structure and mechanical behavior of Al–Al4C3 composites. Russ. Phys. J. **53**, 1153–1157 (2011)
10. Larouche, D.: Computation of solidification paths in multiphase alloys with back diffusion. Calphad **31**, 490–504 (2007)
11. Li, X., Liu, Q.: A version of Hills lemma for Cosserat continuum. Acta Mech. Sin. **25**, 499–506 (2009)
12. Lin, Y.C., Chen, X.-M.: A critical review of experimental results and constitutive descriptions for metals and alloys in hot working. Mater Des. **32**, 1733–1759 (2011)
13. Liu, L., Xu, S., Liu, J., Sun, Z.: Characterization of crystal structure in binary mixtures of latex globules. J. Colloid Int. Sci. **326**, 261–266 (2008)
14. Nembach, E.: Particle Strengthening of Metals and Alloys. Wiley, New York (1997)
15. Nowacki, W.: Theory of Asymmetric Elasticity. Pergamon Press (1986)
16. Parashkevova, L., Bontcheva, N.: Influence of precipitations on elastic plastic properties of Al alloys. Comput. Mater Sci. **47**, 153–161 (2009)
17. Parashkevova, L., Bontcheva, N., Babakov, V.: Modelling of size effects on strengthening of multiphase Al based composites. Comput. Mater Sci. **50**, 527–537 (2010)
18. Peter, I., Rosso, M.: Light alloys from traditional to innovative technologies. In: Ahmad, Z. (ed.) New Trends in Alloy Development, Characterization and Application, Chapter 1, pp. 3–37. InTech (2015)
19. Pierard, O., Friebel, C., Doghri, I.: Mean-field homogenization of multiphase thermo-elastic composites: a general framework and its validation. Compos. Sci. Technol. **64**, 1587–1603 (2004)

Asymptotic Study of the Nonlinear Velocity Problem for the Oscillatory Non-Newtonian Flow in a Straight Channel

Stefan Radev, Sonia Tabakova and Nikolay Kutev

Abstract The studies of non-Newtonian flows, such as blood flows in arteries and polymer flows in channels have very important applications. The non-Newtonian fluid viscosity is modelled by the Carreau model (nonlinear with respect to the viscosity dependence on the shear rate). In the present paper the oscillatory flow of Newtonian and non-Newtonian fluids in a straight channel is studied analytically and numerically. The flow in an infinite straight channel is considered, which leads to a parabolic non-linear equation for the longitudinal velocity. The Newtonian flow velocity is found analytically, while the non-Newtonian velocity is found numerically by the finite-difference Crank-Nicolson method. In parallel, the non-Newtonian (Carreau) velocity is developed in an asymptotic expansion with respect to a small parameter. The zero-th order term of this expansion is exactly the Newtonian velocity solution. The first order term of the velocity expansion is found analytically in terms of higher order harmonics in time. As an example, the polymer solution HEC 0.5% is considered. It is shown that the obtained asymptotic solution and the numerical solution for the non-Newtonian (Carreau) velocity are close for different values of the small parameter.

1 Introduction

The research connected with flows in tubes or channels is very important for medicine, biology, chemical industry, etc. Usually these are flows of blood or some polymer solutions (for example aqueous solution of 0.5% hydroxyethylcellulose, HEC 0.5%), which are considered as shear thinning non-Newtonian fluids. There exist different non-Newtonian rheological models describing the rheology of these fluids.

S. Radev · S. Tabakova (✉)
Institute of Mechanics, BAS, Acad. G. Bontchev str., bl. 4, 1113 Sofia, Bulgaria
e-mail: stabakova@gmail.com

S. Radev
e-mail: stradev@imbm.bas.bg

N. Kutev
Institute of Mathematics and Informatics, BAS, Acad. G. Bontchev str.,
bl. 9, 1113 Sofia, Bulgaria

K. Georgiev et al. (eds.), *Advanced Computing in Industrial Mathematics*, Studies in Computational Intelligence 728,
https://doi.org/10.1007/978-3-319-65530-7_15

The Carreau model [1–6] is one of the most appropriate models for shear thinning fluids, whose viscosities gradually decrease or increase with the shear rate increase or decrease reaching two different plateau values: upper and lower viscosity limits correspondent to small and high shear rates, respectively. If as a reference viscosity the upper limit is considered, then it is possible for some special fluids (for example, some polymer and biological solutions) to construct a small parameter entering in the nonlinear Carreau model function of viscosity.

Often, the flows in tubes or channels have oscillatory or pulsatile character, which complicates significantly the velocity solution search. Only for Newtonian oscillatory flow in straight channels or tubes there exist analytical solutions, respectively for the channel [4–7] and for the tube (the well known solution of Womersley [3, 8]). However, the velocity of the non-Newtonian flow is found only numerically. For the channel it is found by the lattice Boltzman method in [7] and in our previous papers [4–6]—by the finite-difference Crank-Nicolson method. In the last papers we have proved that the solutions, as well as their gradients, are bounded from below and above. We have also proved that the differences between the Newtonian and non-Newtonian (Carreau) velocity and between their gradients on the channel wall are bounded from below and above by constants dependent only on the problem parameters. For some example fluids (blood and HEC 0.5%) it has been shown numerically that the velocity gradient differences on the channel wall between the obtained Newtonian and non-Newtonian (Carreau) solutions are within the theoretically predicted estimates. It occurs that the velocity and the WSS (wall shear stress) of the blood flow, when the blood is considered as Carreau fluid, are close to the velocity and the WSS of the blood when considered as Newtonian fluid with the lower (shear-thinning) viscosity, while for the polymer solution (HEC 0.5%) the velocity and the WSS of the Carreau fluid are close to the velocity and the WSS of the Newtonian fluid with the upper limit viscosity. This fact leads us to the idea to search the velocity solution in an asymptotic expansion about a small parameter that is connected with the physical properties of the considered fluid.

The aim of the present work is to study the non-Newtonian oscillatory flow in a channel, using the Carreau viscosity model: numerically by means of finite-difference Crank-Nicolson method; asymptotically by solution expansion in a small parameter. The zero-th term of this expansion is the Newtonian velocity solution, the first order term is found analytically in terms of higher order harmonics in time. The disposal of this asymptotic solution will be very useful for further theoretical estimates of the difference between the non-Newtonian and Newtonian solutions.

2 Problem Statement

The fluid is assumed with constant density ρ and apparent viscosity μ_{app}, which is constant for the Newtonian model and given by non-linear function of shear rate for the non-Newtonian model. The infinite straight channel is considered in (x, y) space with width $y = H$, where x is the axial coordinate. The pressure p changes only in axial direction with constant gradient, depending only on time: $\frac{\partial p}{\partial x} = -A\cos(nt)$, where A is the oscillation amplitude and n is the angular frequency. These assumptions lead to a single non-zero velocity projection in x direction, namely v_x, which depends only on time t and transverse coordinate y. The equations of motion and continuity [4–6] lead to one single equation for v_x:

$$\rho\frac{\partial v_x}{\partial t} = A\cos nt + \frac{\partial}{\partial y}(\mu_{app}\frac{\partial v_x}{\partial y}) \tag{1}$$

The boundary conditions are $v_x = 0$ at $y = 0$ and $y = H$.

The Carreau viscosity model for μ_{app}, cf. [1] and further denoted as μ_c, is given by:

$$\mu_c = \mu_\infty + (\mu_0 - \mu_\infty)[1 + \lambda^2(\frac{\partial v_x}{\partial y})^2]^{(n_c-1)/2}, \tag{2}$$

where μ_0, μ_∞ are the upper and lower limits of the viscosity corresponding to the low and high shear rates, and λ and n_c are empirically determined for each fluid.

If the following characteristic scales are applied to Eq. (1): H for characteristic length ($y = HY$), $1/n$ for characteristic time ($t = T/n$) and $B = \frac{AH^2}{\mu_0}$ for characteristic velocity ($v_x = Bu$), the dimensionless velocity equation becomes:

$$\alpha_0^2\frac{\partial u}{\partial T} - \frac{\partial}{\partial Y}(\bar{\mu}_{app}\frac{\partial u}{\partial Y}) - \cos(T) = 0, \tag{3}$$

where $\alpha_0 = H\sqrt{\frac{\rho n}{\mu_0}}$ is the Womersley number [8], $\bar{\mu}_{app} = \mu_{app}/\mu_0$ and μ_0 is the fluid viscosity for the fluid assumed as a Newtonian fluid. The dimensionless boundary conditions become:

$$u(T, 0) = u(T, 1) = 0 \tag{4}$$

If the flow is supposed periodic in time T, no initial conditions are prescribed. For the Newtonian flow ($\bar{\mu}_{app} = 1$) the velocity is expected to be a function only of $\cos(T)$ and $\sin(T)$ that will be seen in the next chapter. In the general case of Carreau model, the velocity periodic assumption in time will be discussed further at the numerical solution construction.

3 Results and Discussion

3.1 Newtonian Flow

The solution of the Newtonian fluid flow is denoted by $v(T, Y)$ and Eq. (3) takes the form:

$$8\beta_0^2 v_T - v_{YY} - \cos T = 0 \qquad \text{in} \quad \mathbf{R}\times(0,1) \tag{5}$$

$$v(T,0) = v(T,1) = 0 \qquad \text{for} \quad T \in \mathbf{R}$$

where $\beta_0 = \dfrac{\sqrt{2\alpha_0}}{4}$

The solution of Eq. (5) is given explicitly by [4]

$$v(T,Y) = \frac{1}{8\beta_0^2}[E(Y)\sin T + D(Y)\cos T] \tag{6}$$

where

$$E(Y) = 1 + \frac{S_1(Y)S_2 + C_1(Y)C_2}{S_2^2 + C_2^2} \tag{7}$$

$$D(Y) = \frac{S_1(Y)C_2 - C_1(Y)S_2}{S_2^2 + C_2^2} \tag{8}$$

with

$$S_1(Y) = \sin 2\beta_0(Y - 1/2)\sinh 2\beta_0(Y - 1/2),$$

$$C_1(Y) = \cos 2\beta_0(Y - 1/2)\cosh 2\beta_0(Y - 1/2),$$

$$S_2 = \sin\beta_0 \sinh\beta_0, C_2 = \cos\beta_0\cosh\beta_0.$$

The first and second derivative of velocity $v(T,Y)$ with respect to Y are respectively:

$$v_Y(T,Y) = \frac{1}{4\beta_0(S_2^2 + C_2^2)}\{\sin(T)[CS_1(Y)(C_2 + S_2) - SC_1(Y)(C_2 - S_2)] \tag{9}$$

$$+ \cos(T)[CS_1(Y)(C_2 - S_2) + SC_1(Y)(C_2 + S_2)]\},$$

$$v_{YY}(T,Y) = \frac{1}{S_2^2 + C_2^2}\{\sin(T)[C_1(Y)S_2 - S_1(Y)C_2] + \cos(T)[C_1(Y)C_2 + S_1(Y)S_2]\} \tag{10}$$

with

$$CS_1(Y) = \cos 2\beta_0(Y - 1/2)\sinh 2\beta_0(Y - 1/2),$$

$$SC_1(Y) = \sin 2\beta_0(Y - 1/2)\cosh 2\beta_0(Y - 1/2).$$

Table 1 Parameters for aqueous solution of 0.5% hydroxyethylcellulose (HEC 0.5%)

	λ	μ_0	μ_∞	ρ	n_c
units	s^{-1}	Pas	Pas	kg/m^3	–
HEC 0.5%	0.066	0.22	0.001*	1000	0.5088

∗ viscosity of water

3.2 Carreau Flow

The Eq. (3) for the Carreau viscosity model in dimensionless form is written as:

$$8\beta_0^2 u_T - (\bar{\mu}_c u_Y)_Y - \cos T = 0 \tag{11}$$

where $\bar{\mu}_c = \bar{\mu}_\infty + (1 - \bar{\mu}_\infty)(1 + \bar{\lambda}^2 u_Y^2)^{\frac{n_c-1}{2}}$, $\bar{\mu}_\infty = \dfrac{\mu_\infty}{\mu_0}$, $\bar{\lambda} = \dfrac{\lambda B}{H}$. Together with the conditions (4), Eq. (11) is solved numerically by the Crank-Nicholson method in finite-differences. The Newtonian velocity profile $v(0, Y)$ given by Eq. (6) at $T = 0$ has been chosen as initial condition. The time and space steps are $O(10^{-3})$, ensuring an relative error of $O(10^{-5})$ for the velocity. The calculations have been performed at least for 5 cycles in time ($T = 10\pi$) for achieving convergence to a time-periodic state. Here we can note that the velocity solution does not change, if another initial condition has been used, for example the zero velocity $u(0, Y) = 0$. Then only the convergence time has to be increased, for example up to 10 cycles ($T = 20\pi$).

As an example, the non-Newtonian liquid aqueous solution of 0.5% hydroxyethylcellulose (HEC 0.5%) has been considered, whose physical parameters are summarized in Table 1 [1, 9]. The velocity profiles for different values of $\bar{\lambda}$ are presented in Fig. 1 at $T = 10\pi$ and $\alpha_0 = 0.9256$ ($\beta_0 = 0.3273$). The Newtonian velocity profile corresponds to $\bar{\lambda} = 0$. For small enough values of $\bar{\lambda}$ the Carreau velocity is close to the Newtonian velocity with viscosity μ_0. Similar behaviour of the HEC 0.5% flow velocity has been observed in [6], as well as for the WSS. On contrary, in the case of a blood flow it has been shown in [6] that the Carreau velocity and WSS are closer to the Newtonian velocity and WSS with viscosity μ_∞. This result is very important if one wants to approximate the non-Newtonian fluid flow by a Newtonian flow, i.e., which one of the plateau viscosities to be chosen for the Newtonian fluid flow.

3.3 Asymptotic Analysis of the Carreau Flow

If the parameter $\bar{\lambda} \ll 1$, then it can be chosen as a small parameter $\varepsilon = \bar{\lambda}^2$ in the expression for $\bar{\mu}_c$ and Eq. (11) becomes:

$$8\beta_0^2 u_T - \left[\left(\bar{\mu}_\infty + c(1 + \varepsilon u_Y^2)^{(n_c-1)/2}\right) u_Y\right]_Y - \cos T = 0 \tag{12}$$

where $c = 1 - \bar{\mu}_\infty$.

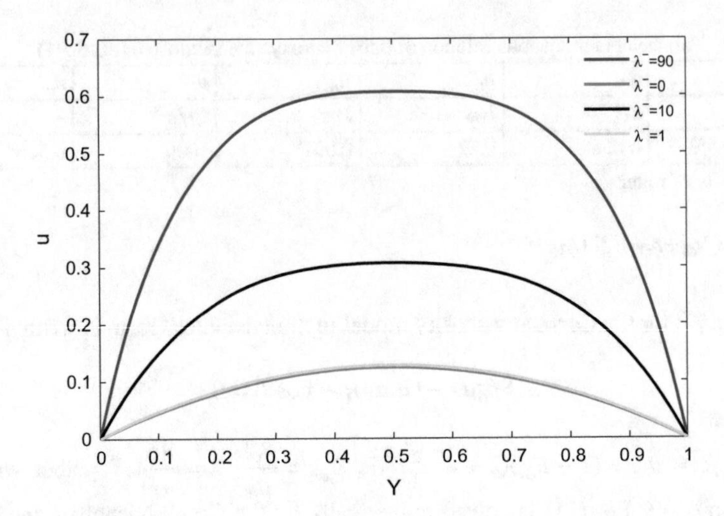

Fig. 1 Comparison between the Newtonian ($\bar{\lambda} = 0$) and Carreau model velocity ($\bar{\lambda} = 90; 10; 1$) for HEC 0.5% at $T = 10\pi$, $\alpha_0 = 0.9256$ ($\beta_0 = 0.3273$)

Next, we shall search the Carreau velocity solution as an asymptotic expansion in ε, i.e.:

$$u(T, Y) = u_0(T, Y) + \varepsilon u_1(T, Y) \ldots \tag{13}$$

For $u_i(T, Y)$ with $i = 0, 1, \ldots$, the Eq. (12) reduces to a linear system of non-homogeneous parabolic differential equations:

$$8\beta_0^2 u_{iT} - u_{iYY} = f_i \tag{14}$$

where $f_0 = \cos T, f_1 = 1.5c(n_c - 1)u_{0Y}^2 u_{0YY}, \ldots$

The solution of Eq. (14) at $i = 0$ is the same as for $v(T, Y)$ given by Eq. (6). Thus, $u_0(T, Y) \equiv v(T, Y)$ and u_{0Y}, u_{0YY} are given respectively by Eqs. (9) and (10).

At $i = 1$, the solution of Eq. (14) is found in the form:

$$u_1(T, Y) = V_{1s}(Y) \sin(T) + V_{1c}(Y) \cos(T) + V_{3s}(Y) \sin(3T) + V_{3c}(Y) \cos(3T), \tag{15}$$

where

$$V_{1s}(Y) = b_4\, C_1(Y) + b_1\, S_1(Y) + \frac{\delta}{10}\left({S_2}^2 + {C_2}^2\right)[\left(-4\, S_2 + 3\, C_2\right) SS_{13}(Y)$$

$$+ \left(4\, S_2 + 3\, C_2\right) SS_{31}(Y) + \left(-3\, S_2 - 4\, C_2\right) CC_{13}(Y) + \left(-3\, S_2 + 4\, C_2\right) CC_{31}(Y)],$$

$$V_{1c}(Y) = \frac{\delta}{10}\left({S_2}^2 + {C_2}^2\right)[-20\, S_2\, S_1(Y) + \left(-3\, S_2 - 4\, C_2\right) SS_{13}(Y)$$

$$+ \left(-3\,S_2 + 4\,C_2\right) SS_{31}(Y) + 20\,C_2\,C_1(Y) + \left(-4\,S_2 - 3\,C_2\right) CC_{31}(Y)$$

$$+ \left(4\,S_2 - 3\,C_2\right) CC_{13}(Y)] - b_1\,C_1(Y) + b_4\,S_1(Y),$$

$$V_{3s} = d_4\,CC_{03}(Y) + d_1\,SS_{03}(Y) + \frac{\delta}{6}[C_2\,\left(3\,S_2{}^2 - C_2{}^2\right)(SS_{33}(Y) + 3\,S_1(Y))$$

$$+S_2\,\left(3\,C_2{}^2 - S_2{}^2\right)\left(3\,C_1(Y) + CC_{33}(Y)\right)],$$

$$V_{3c}(Y) = -\frac{\delta}{6}[\left(-9\,S_2\,C_2{}^2 + 3\,S_2{}^3\right)S_1(Y) + \left(-3\,S_2\,C_2{}^2 + S_2{}^3\right)SS_{33}(Y)$$

$$+ \left(9\,C_2\,S_2{}^2 - 3\,C_2{}^3\right)C_1(Y) + \left(3\,C_2\,S_2{}^2 - C_2{}^3\right)CC_{33}(Y)] - d_1\,CC_{03}(Y) + d_4\,SS_{03}(Y),$$

$$b_1 = -G_2C_2 + G_1S_2, \quad b_4 = G_1C_2 + G_2S_2,$$

$$d_1 = \frac{H_1\,SS_{03}(1) - H_2\,CC_{03}(1)}{CC_{03}^2(1) + SS_{03}^2(1)}, \qquad d_4 = \frac{H_2\,SS_{03}(1) + H_1\,CC_{03}(1)}{CC_{03}^2(1) + SS_{03}^2(1)},$$

$$H_1 = -\frac{\delta}{6}[C_2\,\left(-C_2{}^2 + 3\,S_2{}^2\right)\left(SS_{33}(1) + 3\,S_2\right) + S_2\,\left(3\,C_2{}^2 - S_2{}^2\right)\left(3\,C_2 + CC_{33}(1)\right)],$$

$$H_2 = \frac{\delta}{6}[\left(-9\,S_2\,C_2{}^2 + 3\,S_2{}^3\right)S_2 + \left(-3\,S_2\,C_2{}^2 + S_2{}^3\right)SS_{33}(1)$$

$$+ \left(-3\,C_2{}^3 + 9\,C_2\,S_2{}^2\right)C_2 + \left(-C_2{}^3 + 3\,C_2\,S_2{}^2\right)CC_{33}(1)],$$

$$G_1 = -\frac{\delta}{10}[\left(-4\,S_2 + 3\,C_2\right)SS_{13}(1) + \left(4\,S_2 + 3\,C_2\right)SS_{31}(1)$$

$$+ \left(-3\,S_2 - 4\,C_2\right)CC_{13}(1) + \left(-3\,S_2 + 4\,C_2\right)CC_{31}(1)],$$

$$G_2 = -\frac{\delta}{10}[-20\,S_2\,S_1(1) + \left(-3\,S_2 - 4\,C_2\right)SS_{13}(1) + \left(-3\,S_2 + 4\,C_2\right)SS_{31}(1) + 20\,C_2\,C_1(1)$$

$$+ \left(-4\,S_2 - 3\,C_2\right)CC_{31}(1) + \left(4\,S_2 - 3\,C_2\right)CC_{13}(1)],$$

$$\delta = -\frac{3}{2048}\,\frac{c\left(n_c - 1\right)}{\beta_0{}^4\left(S_2{}^2 + C_2{}^2\right)^3},$$

$$SS_{31}(Y) = \sin 6\beta_0(Y - 1/2)\sinh 2\beta_0(Y - 1/2),$$

$$CC_{31}(Y) = \cos 6\beta_0(Y - 1/2)\cosh 2\beta_0(Y - 1/2),$$

$$SS_{13}(Y) = \sin 2\beta_0(Y - 1/2)\sinh 6\beta_0(Y - 1/2),$$

$$CC_{13}(Y) = \cos 2\beta_0(Y - 1/2)\cosh 6\beta_0(Y - 1/2),$$

$$SS_{03}(Y) = \sin 2\sqrt{3}\beta_0(Y - 1/2)\sinh 2\sqrt{3}\beta_0(Y - 1/2),$$

$$CC_{03}(Y) = \cos 2\sqrt{3}\beta_0(Y - 1/2)\cosh 2\sqrt{3}\beta_0(Y - 1/2),$$

$$SC_{03}(Y) = \sin 2\sqrt{3}\beta_0(Y - 1/2)\cosh 2\sqrt{3}\beta_0(Y - 1/2),$$

$$CS_{03}(Y) = \cos 2\sqrt{3}\beta_0(Y - 1/2)\sinh 2\sqrt{3}\beta_0(Y - 1/2),$$

$$SS_{33}(Y) = \sin 6\beta_0(Y - 1/2)\sinh 6\beta_0(Y - 1/2),$$

$$CC_{33}(Y) = \cos 6\beta_0(Y - 1/2)\cosh 6\beta_0(Y - 1/2).$$

At $i > 1$, the solutions of Eq. (14) contain higher order harmonics in time, but always odd with respect to time, i.e., $\cos((2n + 1)T)$ and $\sin((2n + 1)T)$, where $n > 1$. Thus the velocity solution $u(T, Y)$ will contain an infinite number of odd harmonics in time. This result has been also confirmed after applying the Fourier analysis to a great number of numerical solutions of Eq. (11) for different values of the parameters β_0, $\bar{\lambda}$, n_c and c. The obtained form of the velocity as a periodic function in T confirms our assumption for velocity periodicity.

In order to compare the zero-th order velocity u_0 (Newtonian velocity given by Eq. (6)) with the first order velocity u_1 (Eq. (15)), it is convenient to take the function $\dfrac{u_1}{c(1-n_c)}$, as it depends only on α_0 (or β_0) as well as the velocity u_0 itself. Since the coefficient $0 < c(1 - n_c) < 1$ the function $\dfrac{u_1}{c(1-n_c)}$ is bounded from below by the correspondent values of u_1. In Fig. 2 the maximum values (with respect to time and length) of u_0 and $\dfrac{u_1}{c(1-n_c)}$ are plotted versus α_0. A non-negligible contribution of u_1 to u can be expected only for small α_0 independently on the fluid properties. In all other cases, the higher order terms in the asymptotic expansion will have no contribution. Therefore we confine ourselves only up to the first order term u_1.

The first order velocity u_1 of the HEC 0.5% oscillatory flow is plotted in Fig. 3 for different values of α_0 at time $T = 10\pi$. These velocity profiles are similar in shape with the profiles of the Newtonian velocity u_0, but much smaller in absolute values. With the increase of α_0 the maximum values of u_1 decrease rapidly.

Additionally, we have found that the numerical results for the Carreau velocity of the HEC 0.5% oscillatory flow are very close with the obtained asymptotic solution for different values of the small parameter ε and of α_0. Since their difference is in the range of the numerical method error, we shall not present it.

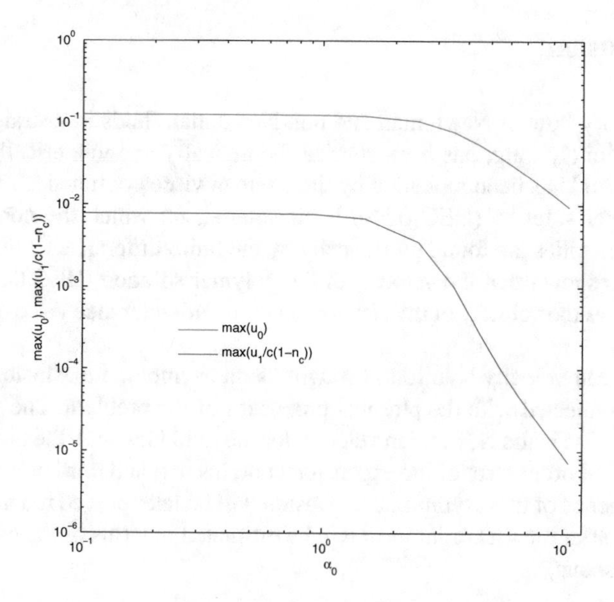

Fig. 2 Maximum zero-th order velocity u_0 (Newtonian velocity given by Eq. (6)) and maximum first order velocity $\dfrac{u_1}{c(1-n_c)}$ of Carreau model as functions of α_0

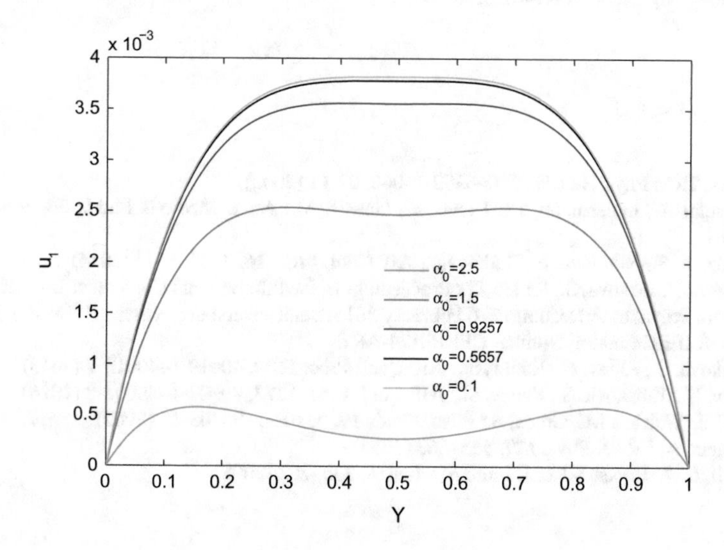

Fig. 3 The first order velocity u_1 of Carreau model for HEC 0.5% as functions of α_0 and Y at $T = 10\pi$

4 Conclusion

The oscillatory flow of Newtonian and non-Newtonian fluids in a straight channel, assumed infinitely long, has been studied theoretically and numerically. The non-Newtonian fluid has been modelled by the Carreau viscosity function. As an example, a polymer solution (HEC 0.5%) is considered, for which the non-Newtonian (Carreau) velocities are found numerically by the finite-difference Crank-Nicholson method. It is shown that the velocity of the polymer solution (HEC 0.5 %) has the same profile as the velocity of the Newtonian fluid with reference viscosity the upper limit one, μ_0.

The Carreau velocity solution is sought as an asymptotic expansion in a small parameter connected with the physical properties of the problem. The zero-th term of this expansion is the Newtonian velocity for the fluid flow with the highest viscosity μ_0. The first order term of the expansion contains first and third order harmonics. The convergence of this asymptotic expansion will be later proved theoretically. The residual cut after the first order term will be estimated in terms of the problem parameters (constants).

Acknowledgements The third author has been partially supported for this research by the National Science Fund of Bulgarian Ministry of Education and Research: Grant DFNI-I02/9 and the other two authors—by Grant DFNI-I02/3.

References

1. Myers, T.G.: Phys. Rev. E **72**, 066302-1–066302-11 (2005)
2. Valencia, A., Ledermann, D., Bravo, E., Galvez, M.: Num. Methods Fluids **58**, 1081–1100 (2008)
3. Tabakova, S., Nikolova, E., Radev, St.:, AIP Conf. Proc. **1629**, 336–343 (2014)
4. Kutev, N., Tabakova, S., Radev, St.: Proceedings of the International Conference on Mechanics-Seventh Polyakhovs Reading, 2–6 February 2015, Saint-Petersburg, Russia, ISBN 978-1-4799-6824-4, IEEE Catalog Number CFP15A24-ART
5. Tabakova, S., Kutev, N., Radev, St.: AIP Conf. Proc. **1690**, 40019-1–40019-7 (2015)
6. Kutev, N., Tabakova, S., Radev, St.: AIP Conf. Proc. **1773**, 80002-1–80002-8 (2016)
7. Boyd, J., Buick, J.M., Green, S.: Phys. Fluids **19**, 93103-1–93103-14 (2007)
8. Womersley, J.R.: J. Phys. **127**, 553–563 (1955)
9. Brush, L.N., Roper, S.M.: J. Fluid Mech. **616**, 235–262 (2008)

Competition for Resources and Space Contributes to the Emergence of Drug Resistance in Cancer

Peter Rashkov

Abstract Recent experiments reveal targeted therapy of tumours promotes the spread of drug-resistant cancer cells in mixed sensitive-resistant tumours. The hypothesis is that drug-stressed sensitive cells produce diffusible growth factors that stimulate the expansion of drug-resistant cells. A mathematical model employing simple ecological competition and a nonlinear motility law is able to reproduce the magnitude of observed expansion of the resistant populations volume without invoking production of diffusible growth factors. The model shows how the therapy-induced removal of the sensitive population alleviates the competitive pressure on the resistant for resources and space and confirms the in vivo experimental findings, and sheds light onto mechanisms behind the large increase of the drug-resistant cancer cells in the treated tumour.

1 Introduction

The in vivo response of tumours with different composition (drug-sensitive cells, drug-resistant cells, and mixed sensitive and resistant cells) to targeted therapy with a variety of kinase inhibitors has recently been studied using wild-type and precultured drug-resistant cell lines of human and mouse melanoma and human lung adenocarcinoma [1]. The effect of therapy on the drug-resistant cells is compared to the control group of *vehicle* treatment with zero inhibitor dose. In [1] it is reported that across all cell lines, targeted therapy of mixed tumours leads to an accelerated proliferation of resistant cells compared to vehicle treatment and proposed that this ensues from a stress response of the drug-sensitive population which release signalling macromolecules (*therapy-induced secretome*) into the tumour microenvironment that support and stimulate the expansion of the drug-resistant population.

Cancer cells respond to stress by releasing diffusible signalling and growth factors into their microenvironment that can potentially stimulate their own or their

P. Rashkov (✉)
Institute of Mathematics and Informatics, Bulgarian Academy of Sciences,
Akad Georgi Bonchev Str., bl. 8, 1113 Sofia, Bulgaria
e-mail: p.rashkov@math.bas.bg

K. Georgiev et al. (eds.), *Advanced Computing in Industrial Mathematics*, Studies in Computational Intelligence 728,
https://doi.org/10.1007/978-3-319-65530-7_16

neighbours' ability to proliferate or acquire resistance [2, 3]. However, there is growing consensus in oncology that cancer growth must be understood from the perspective of the ecosystem interactions within the tumour as well as between the tumour and its microenvironment [4]. In particular, it is quite difficult to quantify the magnitude of a particular factor in vivo among multiple others because of the tumour microenvironment complexity. There is little knowledge of "fundamental population biology parameters" in vivo [5], and even less of the implicit and explicit environmental factors present inside a tumour. While production of diffusible signalling and growth factors can be measured explicitly in vitro, intercellular interactions inside the tumour are difficult to quantify in vivo.

The ecosystem characteristics present in vivo may be concealed, downplayed, distorted or removed, which could lead to an incomplete understanding of the spatio-temporal dynamics inside the tumour, and to misestimation of the role of implicit and explicit factors. An in vivo setting may be characterised by higher interdependence of the different species because of cell-to-cell interaction inside the tissue, competition for resources or space, or cooperation between cells. Furthermore, the in vivo spatial organisation may not be fully reproducible in vitro, but is often omitted from experimental and evolutionary models of cancer [4]. Cells in vitro, for example, may be less restricted in their movement because of the lack of mechanistic obstacles inside the tissue. It is a challenge for a computational model to include the implicit factors' effect on tumour evolution, and development of therapy.

Phenotypes abound in ecosystems due to various metabolic and physiological trade-offs [6], and cancer is no exception. Drug resistance may lead to various trade-offs: metabolism and decreased proliferation rate [5], or higher vulnerability in competition for resources. Thus, the fitness cost of drug resistance should be considered in models of tumour evolution.

In an ecosystem where populations with various characteristics compete for limited resources, a reduction in one of the populations may trigger great changes in the abundance of competitors. Such an implicit proliferative advantage could be quite substantial, as the mixed tumours used by [1] contain pre-cultured resistant cells which are immune to therapy. By eliminating the drug-sensitive population, the therapy could in effect liberate the drug-resistant population from the competitive pressure for resources and space and create a more favourable environment for its expansion.

In order to understand the scale of environmental factors behind the promotion of drug resistance under therapy of a tumour a mathematical model based on ecological principles of competition is proposed to simulate in silico the experiment of [1] and compare the model predictions to the experimental observations. The model addresses the questions about

- the growth advantage of the resistant population in mixed tumours caused by reduced competition for space and resources,
- the dynamics of volume of resistant tumours in mixed tumours subject to therapy or vehicle treatment,

- the effect of the density-dependent cell motility in the evolution of tumour composition.

The model demonstrates that therapy can accelerate growth of the drug-resistant population even when any effect of therapy-induced secretomes is neglected. Under certain conditions the model predicts that in a treated mixed tumour the volume of resistant population increases faster compared to an identical vehicle-treated tumour, and suggests a substantial therapy-induced colonisation effect, even when the resistant population is at a proliferative disadvantage compared to the sensitive due to costs of resistance. The model predictions are in accordance with preliminary in vitro data on growth of drug-resistant lung carcinoma cells in mixed tumours [7], where the growth of the resistant cells is suppressed in presence of sensitive cells. The model raises awareness of the importance of implicit ecosystem interactions which might be difficult to measure or reproduce explicitly in vitro.

2 Mathematical Model

The model uses a continuous approach describing the spatio-temporal evolution of cell population densities and concentrations of chemicals. The spatial domain Ω describes the tumour and its vicinity, and in this study Ω will be either $\mathbb{R}$ or a convex domain $\Omega \subset \mathbb{R}^2$ with piecewise smooth boundary. The index $i = 0, 1, 2$ refers to the population of healthy, drug-sensitive and drug-resistant cells. $u_i(t, x)$ denotes the i-th population density at time $t > 0$ at point $x \in \Omega$.

Cancer cells upregulate the glycolitic metabolic pathway which leads to increased production of lactic acid [8], the excess of which is then secreted into the extracellular space. Increased pH of the tumour microenvironment above the homeostatic level decreases the viability of the surrounding healthy tissue via acidosis [9–11]. Lactic acid $v(t, x)$ is secreted proportionally to total cancer cell density $u_1 + u_2$ at rate $r > 0$ and reabsorbed via the tissue at rate $\kappa_v > 0$. The free acid diffuses in Ω with diffusion rate D_v.

For simplicity the drug $w(t, x)$ is administered at a constant rate $W_0 > 0$, which is a good approximation to the experiment [1] (drugs administered once or twice daily). Drug uptake by the cancer population i is modelled by Michaelis-Menten terms with maximal uptake rates V_i and Michaelis constants $k_i^M, i = 1, 2$. The drug diffuses in Ω with a rate $D_w > 0$ and is reabsorbed at a rate κ_w. Metabolism with Michaelis-Menten kinetics is used in other models of tumour growth [12–14], which, however, employ agent-based modelling for the cells.

Cell population i follows a logistic growth law with proliferation rate $g_i > 0$, carrying capacity K and Lotka-Volterra parameters σ_{ij} for the competition from population j. In addition, there is an excess death term for the healthy cells due to acidosis, which is linear in v with parameter $d_0 > 0$. Furthermore, the drug's pro-apoptotic effect is modelled by an excess death term for the cancer populations, equal to the respective uptake rate of the drug. In other words, no drug is lost during the uptake into the cell and the drug's pro-apoptotic efficiency amounts to 100%. Resistant cells normally survive higher inhibitor doses, so the model assumes $V_1 \gg V_2$ and $k_1 \ll k_2$. The drug does not affect the surrounding healthy tissue.

Both types of cancer cells are assumed to have a proliferative advantage over the healthy cells [15], so the intrinsic population proliferation rates satisfy $g_0 < g_1, g_2$. The fitness cost of resistance [16, 17] is encoded in g_i and the Lotka-Volterra competition parameters σ_{ij}, so $g_1 > g_2$ [18, 19]. A value $\sigma_{10} = 0$ means that the healthy cells do not affect the proliferation of the sensitive cancer cells, while $\sigma_{20} > 0$ amounts to a competition effect of the healthy cells on the resistant cells (stemming from a physiological trade-off between resistance and vulnerability in the resistant cells). Values $\sigma_{12}, \sigma_{21} > 0$ account for competition for resource between the both cancer population types.

Tumours exhibit heterogeneous growth possibly due to mechanistic effects (such as crowding) at the interface between tissues of different composition [20]. The competition for space between the cells at the tumour-healthy tissue as well as inside the tumour interface is captured by a nonlinear term for cell motility. Cell motility is modelled following [21, 22], where in the presence of multiple competing cell populations $u_i, i = 0, 1, 2$ at the same point in Ω, the flux for the i-th population $\mathbb{J}_i$ is proportional to the total flux $-\nabla(u_0 + u_1 + u_2)$. The contribution of i-th population to the overall cell flux is density-dependent according to the fraction of the population u_i inside the total population,

$$\mathbb{J}_i(u_0, u_1, u_2) = -\frac{u_i}{u_0 + u_1 + u_2}(\nabla u_0 + \nabla u_1 + \nabla u_2). \tag{1}$$

The motility term for population i is the divergence of the flux, $\nabla \cdot \mathbb{J}_i$.

Combining these considerations gives the governing equations for the population densities and chemical concentrations

$$\frac{\partial u_0}{\partial t} = \nabla \cdot \mathbb{J}_0 + g_0 u_0 \left(1 - \frac{u_0}{K} - \sigma_{01} u_1 - \sigma_{02} u_2\right) - d_0 v u_0, \tag{2}$$

$$\frac{\partial u_1}{\partial t} = \nabla \cdot \mathbb{J}_1 + g_1 u_1 \left(1 - \frac{u_1}{K} - \sigma_{12} u_2\right) - \frac{V_1 w}{k_1^M + w} u_1, \tag{3}$$

$$\frac{\partial u_2}{\partial t} = \nabla \cdot \mathbb{J}_2 + g_2 u_2 \left(1 - \sigma_{20} u_0 - \sigma_{21} u_1 - \frac{u_2}{K}\right) - \frac{V_2 w}{k_2^M + w} u_2, \tag{4}$$

$$\frac{\partial v}{\partial t} = D_v \nabla^2 v + r(u_1 + u_2) - \kappa_v v, \tag{5}$$

$$\frac{\partial w}{\partial t} = D_w \nabla^2 w + W_0 - \frac{V_1 w}{k_1^M + w} u_1 - \frac{V_2 w}{k_2^M + w} u_2 - \kappa_w w. \tag{6}$$

Setting $u_i = u_i/K, w = w/K, c_{ij} = K\sigma_{ij}, \rho = Kr, w_0 = W_0/K, k_i = k_i^M/K$ the above system becomes non-dimensionalised,

$$\frac{\partial u_0}{\partial t} = \nabla \cdot \mathbb{J}_0 + g_0 u_0(1 - u_0 - c_{01}u_1 - c_{02}u_2) - d_0 v u_0, \tag{7}$$

$$\frac{\partial u_1}{\partial t} = \nabla \cdot \mathbb{J}_1 + g_1 u_1(1 - u_1 - c_{12}u_2) - \frac{V_1 w}{k_1 + w} u_1, \tag{8}$$

$$\frac{\partial u_2}{\partial t} = \nabla \cdot \mathbb{J}_2 + g_2 u_2(1 - c_{20}u_0 - c_{21}u_1 - u_2) - \frac{V_2 w}{k_2 + w} u_2, \tag{9}$$

$$\frac{\partial v}{\partial t} = D_v \nabla^2 v + \rho(u_1 + u_2) - \kappa_v v, \tag{10}$$

$$\frac{\partial w}{\partial t} = D_w \nabla^2 w + w_0 - \frac{V_1 w}{k_1 + w} u_1 - \frac{V_2 w}{k_2 + w} u_2 - \kappa_w w. \tag{11}$$

In Eqs. 7–11 the unknowns u_i, v, w are subject to homogeneous Neumann boundary conditions on $\partial\Omega$.

Reductions of this model employing only cell populations allow for abundance of solutions, not only travelling waves of cells, but also stable gradients between the competing cell populations instead of perfect mixing [21, 23–25].

3 Results

Equations 7–11 form a strongly coupled parabolic system with nonlinear diffusion terms. To shed light onto the speed of the propagating wave front of drug-resistant cells, and hence onto the increase in volume of the drug-resistant population inside the tumour, two subproblems of pairwise interactions will be considered over $\mathbb{R}$:

P1. resistant versus sensitive cancer populations with therapy and vehicle treatment ($u_0 \equiv 0, v \equiv 0$ for all $x \in \mathbb{R}, t > 0$). This case describes the dynamics of competition inside the tumour.

P2. cancer against healthy tissue (either $u_1 \equiv 0$ or $u_2 \equiv 0$ for all $x \in \mathbb{R}, t > 0$). This case describes the dynamics at the interface between healthy tissue and invasive (either sensitive or resistant) cancer.

P1 and P2 are analysed for multistationarity and existence of travelling wave solutions, whose their wave speed is computed numerically to elucidate the dynamics of the full problem.

With regard to P1, if $w_0 > 0$ is large enough so that the sensitive cancer cells become extinct (in other words, $\limsup_{t\to\infty} u_1(t, x) = 0$), the wave front of u_2 converges to a wave front whose profile and speed are approximated by the Fisher-KPP equation. Also for $w_0 \equiv 0$ the reaction part of Eqs. 8–9 allows several equilibria $(\hat{u}_1, \hat{u}_2)$: a full extinction state $E_0 = (0, 0)$, two mutually exclusive states $E_1 = (0, 1), E_2 = (1, 0)$, and for appropriately chosen competition parameters c_{ij}, a coexistence state exists, $E_3 = (\frac{1-c_{12}}{1-c_{12}c_{21}}, \frac{1-c_{21}}{1-c_{12}c_{21}})$ [26]. Furthermore, E_0 is unconditionally unstable, and the local stability of E_1, E_2 depends on the parameters as follows: if $c_{21} < 1$, E_1 is a saddle, if $c_{12} < 1$, E_2 is a saddle, and if $c_{12}, c_{21} < 1$, E_3 exists and is locally stable. In a physiologically relevant range, the competitions parameters have to satisfy $c_{12}, c_{21} < 1$, to ensure the coexistence of both types of cancer cells inside the tumour, and the coexistence state E_3 is locally stable.

P2 describes the wave of invasive cancer u_j against healthy cells u_0. If $w_0 > 0$ is large enough so that $\limsup_{t\to\infty} u_j(t,x) = 0$ (the cancer population becomes extinct), the wave front of u_0 converges to a wave front whose profile and speed are again approximated by the Fisher-KPP equation. The local stability of spatially homogeneous equilibria $(\hat{u}_0, \hat{u}_j, \hat{v})$ of P2 for $w_0 \equiv 0$ is easily determined. The following equilibria are physiologically relevant, $E_1 = (1,0,0)$ corresponding to homeostasis (prevalence of healthy tissue) and $E_2 = (0, 1, \frac{\rho}{\kappa_v})$ corresponding to tumour. Analysis of the eigenvalues of the Jacobians in E_1 and E_2 gives the following: whenever $c_{j0} < 1$, E_1 is locally unstable. If $g_0(1 - c_{0j}) - d_0\frac{\rho}{\kappa_v} < 0$, E_2 is locally stable. This is consistent with the choice of values $c_{10}, c_{20} < 1$ and $c_{01}, c_{02} > 1$, so that on one hand, E_1 is locally unstable to perturbations due to the presence of cancer cells, and on other hand, E_2 is locally stable if either the secretion rate of lactic acid ρ or the excess acidification death rate d_0 are large enough. Furthermore, such values preclude coexistence between either type of cancer and healthy cells.

3.1 Travelling Wave Solutions

Numerical simulations over a finite one-dimensional domain help estimate the profile of the travelling wave front and the wave speed of P1 and P2 for vehicle treatment ($w_0 = 0$). Model parameters are listed in Table 1 unless indicated otherwise. Initial data is chosen with separated populations of the two cell types on the interval $(-100, 100)$. Without loss of generality it is assumed henceforth that $c_{12} = c_{21} := c$ (values in Table 1). The initial conditions for P1 are

$$u_1(0,x) = 1 - \frac{c}{1+c}(1+e^{\xi x})^{-1}, \quad u_2(0,x) = \frac{(1+e^{\xi x})^{-1}}{1+c} \tag{12}$$

to give two overlapping cancer cell populations for $x < 0$, and rapid convergence $u_1(0,x) \to 1, u_2(0,x) \to 0$ for $x \to +\infty$. The initial conditions for P2 are

$$u_0(0,x) = (1+e^{\xi(x_0-x)})^{-1}, \quad u_j(0,x) = (1+e^{\xi(x-x_*)})^{-1}, \quad v(0,x) = 0. \tag{13}$$

with $x_0 > x_*$ to give two separated populations which overlap at the interface between the cancer population u_j and the healthy tissue u_0. Larger values of $\xi > 0$ correspond to faster decay of the initial conditions, and highly localised initial cell populations.

The following observations describe qualitatively the behaviour of the propagating wave fronts. It is expected that the decay of the travelling wave solution will depend on the decay of the initial data, ξ. For lower ξ (corresponding to faster decay of initial data), the wave speed a will be higher [21]. The numerical analysis of travelling wave solutions for P1 confirms this: a ranges between 0.012 and

Table 1 Parameter values used in the simulations

Parameter	Meaning	Value	Reference
g_0	Intrinsic proliferation rate healthy cells	$0.025\ \text{h}^{-1}$	[27]
g_1	Intrinsic proliferation rate drug-sensitive cancer cells	$0.050\ \text{h}^{-1}$	[18]
g_2	Intrinsic proliferation rate drug-resistant cancer cells	$0.030\ \text{h}^{-1}$	[18]
K	Carrying capacity all cells	$5 \cdot 10^4 \text{mm}^{-3}$	[28]
V_1	Maximum uptake drug drug-sensitive cancer cells	$0.75\ \text{h}^{-1}$	Estimate
k_1	Michaelis constant drug-sensitive cancer cells	$3\ \mu\text{Mmm}^3$	Estimate
V_2	Maximum uptake drug drug-resistant cancer cells	$0.375\ \text{h}^{-1}$	Estimate
k_2	Michaelis constant drug-resistant cancer cells	$200\ \mu\text{Mmm}^3$	Estimate
d_0	Death rate due to acidity	$0.04\ \mu\text{M}^{-1}\text{h}^{-1}$	[28]
ρ	Lactic acid production rate	$4\ \mu\text{Mh}^{-1}$	[28]
κ_v	Reabsorption rate lactic acid	$0.396\ \text{h}^{-1}$	[28]
κ_w	Reabsorption rate drug	$0.07\ \text{h}^{-1}$	Estimate
w_0	Administration rate drug	As shown	
D_v	Diffusion rate lactic acid	$2\ \text{mm}^2\text{h}^{-1}$	[28]
D_w	Diffusion rate drug	$1.2\ \text{mm}^2\text{h}^{-1}$	Estimate

Lotka-Volterra competition matrix

c_{ij}	0	1	2
0	–	0.9	0.9
1	0	–	0.9
2	0.9	0.9	–

0.006 as ξ ranges between 0.3 and 1.5. In P1, if the sensitive population were removed by administering at dose $w_0 > 0$ so that $\limsup_{t\to\infty} u_1(x,t) = 0$, the dynamics of u_2 is approximated by the Fisher-KPP equation with minimal wave speed $a_{min} = 2\sqrt{g_2} \approx 0.3464$. Thus, the nonlinear motility law (Eq. 1) significantly reduces the travelling wave speed.

Furthermore, for a fixed ξ the wave speed a in P2 will depend on the secretion rate of lactic acid ρ: the higher ρ, the higher the excess acidification effect on the healthy tissue. Hence, at the interface between cancer and healthy cells, the latter will be removed faster, opening up space for colonisation. Figure 1a plots the computed wave speed a in the range $\rho \in (0, 0.2)$ for the invading wave of cancer (data is plotted for both sensitive and resistant populations).

Considering the sensitive population in P2 ($u_2 \equiv 0$) the reaction term in Eq. 8 depends only on u_1. Hence, if cell motility were modelled by linear diffusion, Eq. 8 would be approximated by the Fisher-KPP equation with proliferation rate g_1. As u_1 is effectively decoupled from u_0, the minimal wave speed can be computed

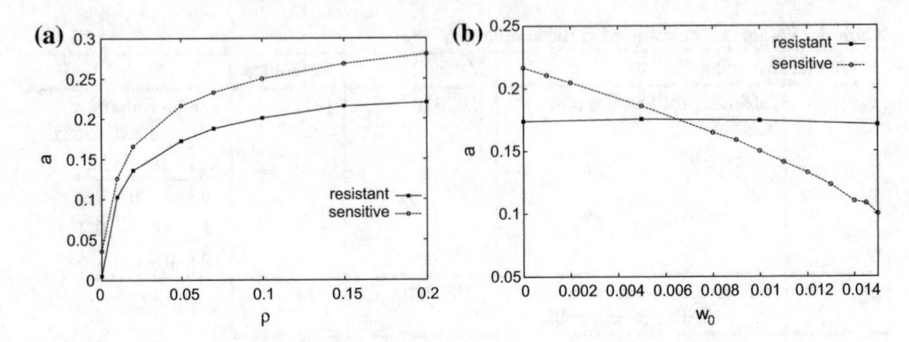

Fig. 1 **a** Computed wave speed a for the invasive cancer wave (P2) as function of ρ, with $\xi = 1.5$. **b** Computed wave speed a for the invasive cancer wave P2 as function of w_0 (the decay parameter of initial data $\xi = 1.5$ in Eqs. 12 and 13

directly [29], and equals $a_{min} = 2\sqrt{g_1} \approx 0.4472$ which exceeds the numerically computed values when $w_0 = 0$ (Fig. 1).

Figure 1b plots the computed travelling wave wave speed in P2 as a function of the drug administration rate w_0. The simulations show that the invasive spread of cancer slows down in the case of sensitive population, because the net growth rate is reduced by the therapy. The speed of spread for resistant population is basically unchanged because they are less responsive to the drug.

The results from this numerical study of the pairwise dynamics of the speed of a propagating wave of drug-resistant cancer cells inside the tumour and against the healthy cells in one spatial dimension reveal fundamental differences in dynamics. The resistant cells expand at a significantly lower speed inside the tumour compared to the healthy tissue (or a free edge of no cells). In a higher dimensional setting, thus, it is expected that the growth of the resistant population will be suppressed if it doesn't have a boundary towards the healthy tissue or towards a free edge. If therapy creates a free edge by removing the sensitive population in a vicinity of the resistant, the latter will grow in this direction.

3.2 *Tumour Volume Expansion*

Knowledge of the qualitative behaviour from the one-dimensional analysis and simulations can give insight into the behaviour of the dynamical system in higher dimensions. The temporal evolution of the volume of the sensitive and resistant populations in mixed tumours is studied for vehicle and therapy. For a given two-dimensional domain Ω, the volumes at time $t > 0$ are given by

$$V_{\text{sen}}(t) = \int_{\Omega} u_1(t, \cdot), \quad V_{\text{res}}(t) = \int_{\Omega} u_2(t, \cdot). \tag{14}$$

Simulations are performed on a square domain $\Omega = (0, 90)^2$ with three sets of initial conditions of cancer cell distributions, denoted by Roman numerals. The initial conditions $u_1(0, \cdot), u_2(0, \cdot)$ are smooth and rapidly decaying functions over Ω, with maxima over compact sets of different area and localisation in Ω. For the simulations the bulk of tumour cells is localised at the interior of Ω sufficiently far from $\partial\Omega$ and surrounded by healthy tissue. In more detail, the initial conditions are

I. $u_1(x, y) = [(1 + e^{\xi(40-x)})(1 + e^{\xi(x-65)})(1 + e^{\xi(y-55)})(1 + e^{\xi(30-y)})]^{-1}$,
$u_2(x, y) = [(1 + e^{\xi(35-x)})(1 + e^{\xi(x-45)})(1 + e^{\xi(y-50)})(1 + e^{\xi(40-y)})]^{-1}$,
$u_0 = (1 - u_1)(1 - u_2)$.

II. $u_1(x, y) = 0.9[(1 + e^{\xi(35-x)})(1 + e^{\xi(x-45)})(1 + e^{\xi(y-45)})(1 + e^{\xi(35-y)})]^{-1}, \quad u_2 = u_1$,
$u_0 = 1 - u_1/0.9$.

III. $u_1(x, y) = [(1 + e^{\xi(25-x)})(1 + e^{\xi(x-55)})(1 + e^{\xi(y-55)})(1 + e^{\xi(25-y)})]^{-1}, u_2(x, y) = [(1 + e^{\xi(35-x)})(1 + e^{\xi(x-45)})(1 + e^{\xi(y-45)})(1 + e^{\xi(35-y)})]^{-1}, u_0 = 1 - u_1$.

The parameter ξ determines the rate of decay of the initial conditions, and for the simulations $\xi = 1.5$ is chosen to localise the tumour at $t = 0$.

Observe that in the simulation the ratio $V_{res}(0)/V_{sen}(0)$ at the start of therapy varies from 0.1 (set I) to 1 (set II). In the experimental setup [1] the mice are inoculated with much lower ratio ($V_{res}/V_{sen} \approx 0.0005$), but then the tumours are allowed to establish before the start of therapy, so the actual ratio $V_{res}(0)/V_{sen}(0)$ is unknown.

The simulation is performed over a time interval such that the peak of the cancer cell densities remains strictly separated from Ω in order to avoid any refractory waves of u_1, u_2, v, w which might arise due to the homogeneous Neumann boundary conditions on $\partial\Omega$. The simulations of *vehicle* treatment are run with value $w_0 = 0$.

Figure 2a displays histograms of the relative change $V_{res}(100)/V_{res}(0)$ for vehicle and for therapies with different w_0. Figure 2b displays the results for the increase in $V_{res}(100|therapy)$ for case III normalised so that $V_{res}(100|vehicle) \equiv 1$.

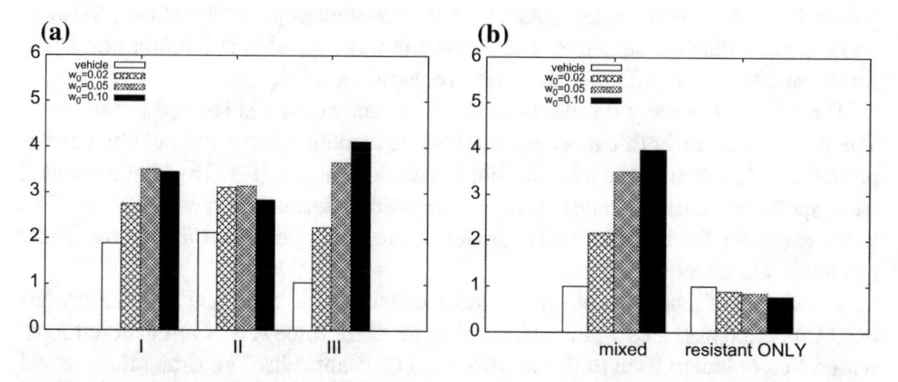

Fig. 2 **a** Relative change in the volume of resistant population ($V_{res}(100)/V_{res}(0)$) inside a mixed tumour for vehicle and therapy with different administration rates w_0 for the three sets of initial conditions. **b** Changes in volume of $V_{res}(100|therapy)$ relative to $V_{res}(100|vehicle)$ for two tumours with equal $V_{res}(0)$ but of different initial composition: a mixed (sensitive + resistant) tumour (*left*) and a fully resistant tumour (*right*). Simulations for the mixed tumour use set III of initial conditions

4 Discussion

A major obstacle for cancer therapy is the emergence of cell clones inside the tumour which are resistant to therapy, eventually leading to tumour relapse in the medium or long term. Cancer development involves a multitude of physiological, environmental and evolutionary factors whose interaction must be well understood in order to develop successful strategies for control and cure of the disease [15].

Proliferation rates of cells used in the experiment in [1] were surveyed to set the model parameters in a biologically relevant range. Evidence in support of trade-offs between resistance and proliferation rate was gathered from published data [18, 19]. Quantitative data on trade-offs between viability in resource competition and resistance could not be found. The experiments of [1] are done with several cancer cell lines whose proliferation rates could be characterised only approximately using available data [18, 27, 30]. Healthy cells are modelled with a lower proliferation rate compared to the cancer cells, $g_0 < g_2 < g_1$. Furthermore, the proliferation rate of the resistant population used in the model is assumed lower than the estimate for the resistant PC-9/ER3 cell line [18, Fig. 1C].

To study the effect of therapy on an idealised tumour microenvironment consisting of sensitive and resistant cells, let us discuss the two subproblems that have been studied numerically to obtain understanding of the full model.

The model accounts for the aggressiveness/invasiveness of the cancer population via the production rate of lactic acid ρ, as a relation between lactate levels in a tumour and its metastatic potential is postulated [31]. Since the model assumes a physiological trade-off for the resistant population leading to a lower proliferation rate than sensitive ($g_2 < g_1$), the wave speed of the sensitive cell wave front against the healthy cells is greater than that of the resistant cell front for each $\rho > 0$. Numerical computations of the wave speed for invasive cancer over $\mathbb{R}$ confirm these observations because that the wave speed of invading cancer increases monotonically with ρ (Fig. 1b). The numerical analysis shows the resistant population exhibits a slower wave speed a than the sensitive. For $\rho = 0$, the wave speed is at a minimum: 0.005 for the resistant, and 0.036 for the sensitive population (Fig. 1b).

The effect of therapy on the wave speed of cancer cells is studied numerically. The wave speed of both cancer populations advancing against the healthy cells is plotted as a function of the inhibitor administration rate w_0 (Fig. 1b). The computed wave speed for sensitive cancer cells is a monotone decreasing function of w_0. The wave speed for the resistant cells remains more or less constant due to the lesser pro-apoptotic effect.

The spread of resistant cells in a tumour under vehicle treatment and therapy has been simulated over a two-dimensional domain Ω for three sets of initial conditions, with different localisation of the sensitive and resistant cells. The simulations reveal that the dynamics of $V_{\text{res}}(t)$ are quite different across the three cases and depend strongly on the initial conditions. For initial data III the resistant cells are initially surrounded by sensitive cells within the tumour. When treated with vehicle ($w_0 \equiv 0$), their spatial spread is highly inhibited by the presence of sensitive cells. This outcome is corroborated by the computed wave speed of the resistant population in the one-dimensional simulation (Fig. 1) and experimental data on suppressed growth of the resistant population inside a mixed sensitive-resistant cancer cell culture [7]. The travelling wave solutions of P1 are characterised by low wave speed. In the two-dimensional simulation V_{res} increases by 4% between $t = 0$ and $t = 100$ hours for vehicle treatment (Column III in Fig. 2). Under therapy the sensitive population declines ($\lim_{t\to\infty} V_{sen}(t) = 0$), alleviating the competition pressure on the resistant population, leading to a four-fold increase in V_{res} over the same period.

For mixed tumours with initial data in cases I and II, the maxima of $u_2(0, \cdot)$ border initially partially or fully onto healthy cells, and, therefore, the resistant population is able to expand in that direction at a rate which is very weakly dependent of the application of therapy (compare the computed wave speed in Fig. 1b) and is confirmed by the simulation results For case I, the ratio $2.7 < V_{res}(100)/V_{res}(0) < 3.5$ for therapy is not much different from that for vehicle treatment (≈ 2.5) (Column I in Fig. 2a). For case II, the resistant population initially has its entire free boundary towards the healthy cells, so the ratios $V_{res}(100)/V_{res}(0)$ for therapy and for vehicle are of the same magnitude (Column II in Fig. 2a).

The model is able to reproduce qualitatively in silico the in vivo observations, but it has certain limitations. The computational model and the experimental setup assume that resistant cells are intrinsically present inside the rumour, and so are immune to therapy. To what extent this phenomenon occurs in in situ tumours is a matter of debate [32, 33]. The model assumes for simplicity continuous infusion of inhibitor which is an idealised approach for computational simplicity. Finally, the organism's immune response is neglected. Nevertheless, the model strives to focus attention that an ecosystem model of a spatially-distributed tumour microenvironment is able to reproduce qualitatively experimental results, and bring experimentalists' attention to physiological trade-offs and the cost of resistance which are often neglected in cancer studies.

5 Conclusion

This study considers an ecosystem model of a spatially-distributed tumour microenvironment subject to therapy with a proapoptotic drug. The growth of three cell populations (healthy cells, drug-sensitive and drug-resistant cancer cells) is based on a logistic model with Lotka-Volterra-type competition, supplemented by two types of interaction due to presence of diffusible chemicals (drug administered during treatment, and lactic acid secreted by the cancer cells). Physiological trade-offs and the cost of resistance are incorporated into the model parameters. The cell movement is based on the assumption that cells are restricted mechanistically by other cells in their vicinity, leading to a crowding effect [20], and represented by a density-dependent diffusion term [21, 22].

The model strives to answer the question whether a therapy-induced alleviation of environmental competition and a density-dependent cell motility can explain the magnitude of experimentally observed growth of the resistant population in mixed sensitive-resistant tumours, without invoking production of growth factors as a stress response. Numerical simulations show that environmental competition in a mixed tumour can deter the spread and proliferation of a resistant population which exhibits a proliferative disadvantage due to costs of resistance, and agree with preliminary experimental data on co-cultured sensitive and resistant cells [7]. Administration of drugs targeting sensitive cells, in fact, can alleviate intratumour environmental competition and cause significant growth advantage for drug-resistant cells in mixed tumours. Their volume can grow substantially inside a treated mixed tumour compared to an identical untreated tumour, and the relative difference resembles qualitatively that in experimental data [1, Suppl. Fig. 1g, i, j, k].

Furthermore, density-dependent cell motility is an important factor for the changes of volume of sensitive or resistant populations in a mixed tumour. The simulation results across different initial data demonstrate that the spatial structure of the tumour plays an important role in the outcome. Tumours with similar ratios of sensitive to resistant cells may exhibit different response to treatment because of different spatial structure. Unfortunately the role of tumour spatial structure and organisation remain underestimated in experimental and evolutionary models of cancer [4]. The simple model proposed here shows how the environmental factors present in vivo may combine with density-dependent cell motility to influence the dynamics of tumour volumes. More experimental data on interaction between sensitive and resistant cancer cell lines in vitro could bring insights for therapy design, especially for containment of resistance.

Acknowledgements The author acknowledges funding for his postdoc position at the University of Exeter from AstraZeneca (Cambridge, UK). Thanks to Mark Hewlett and Bogna Pawłowska for proofreading the manuscript.

Appendix

Here is a brief summary of the numerical method used to solve Eqs. 7–11. The main challenge is the treatment of the non-linear flux term. For the 1D problems P1 and P2, the flux term is to rewritten as in [21]

$$\frac{\partial}{\partial x}\left(\frac{u}{u+v}\frac{\partial}{\partial x}(u+v)\right) = \frac{u}{u+v}\cdot\frac{\partial^2}{\partial x^2}(u+v) + \frac{\partial}{\partial x}(u+v)\cdot\frac{\partial}{\partial x}\left(\frac{u}{u+v}\right) \tag{15}$$

and then the equations are integrated in time using the ROWMAP solver [34].

For the 2D problem, a variational formulation scheme is used to discretise the equations in space with Lagrangian P2 finite elements. The non-linear system is integrated in the software environment FreeFem++ [35] by a fully implicit Euler scheme, and the equations are solved iteratively at every time step by Newton's method. For this purpose, the non-linear diffusion term is approximated as follows

for a test function $\psi \in H^1(\Omega)$. The divergence theorem employing the homogeneous Neumann boundary conditions for u_j gives

$$\int_\Omega \psi \nabla \cdot \left(\frac{u_j(\sum_{i=0}^2 \nabla u_i)}{\sum_{i=0}^2 u_i} \right) dx = - \int_\Omega \frac{u_j(\sum_{i=0}^2 \nabla u_i)}{\sum_{i=0}^2 u_i} \nabla \psi \, dx := - \int_\Omega F_j \nabla \psi \, dx. \tag{16}$$

In order to use Newton's method for the solution of Eqs. 7–11, F_j in Eq. 16 is expanded in a Taylor series. For a small perturbation $\tilde{u}_i$ in u_i, F_0, for example, is

$$F_0(u_0 + \tilde{u}_0, u_1 + \tilde{u}_1, u_2 + \tilde{u}_2) \approx \frac{(u_1 + u_2)(\sum_{i=0}^2 \nabla u_i)\tilde{u}_0}{(\sum_{i=0}^2 u_i)^2} - \frac{u_0(\sum_{i=0}^2 \nabla u_i)\tilde{u}_1}{(\sum_{i=0}^2 u_i)^2}$$
$$- \frac{u_0(\sum_{i=0}^2 \nabla u_i)\tilde{u}_2}{(\sum_{i=0}^2 u_i)^2} + \frac{u_0(\sum_{i=0}^2 \nabla \tilde{u}_i)}{\sum_{i=0}^2 u_i} + \text{higher order terms.}$$

A similar approximation is made for the other two terms F_1, F_2.

References

1. Obenauf, A., Zou, Y., Ji, A., Vanharanta, S., Shu, W., Shi, H., Kong, X., Bosenberg, M., Wiesner, T., Rosen, N., Lo, R., Massague, J.: Therapy-induced tumour secretomes promote resistance and tumour progression. Nature **520**, 368–372 (2015)
2. Sun, Y., Campisi, J., Higano, C., Beer, T., Porter, P., Coleman, I., True, L., Nelson, P.: Treatment-induced damage to the tumor microenvironment promotes prostate cancer therapy resistance through WNT16B. Nat. Med. **18**, 1359–1368 (2012)
3. Wilson, T., Fridlyand, J., Yan, Y., Penuel, E., Burton, L., Chan, E., Peng, J., Lin, E., Wang, Y., Sosman, J., Ribas, A., Li, J., Moffat, J., Sutherlin, D., Koeppen, H., Merchant, M., Neve, R., Settleman, J.: Widespread potential for growth-factor-driven resistance to anticancer kinase inhibitors. Nature **487**, 505–509 (2012)
4. Korolev, K., Xavier, J., Gore, J.: Turning ecology and evolution against cancer. Nat. Rev. Cancer **14**, 371–380 (2014)
5. Aktipis, C., Boddy, A., Gatenby, R., Brown, J., Maley, C.: Life history trade-offs in cancer evolution. Nat. Rev. Cancer **13**, 883–892 (2013)
6. Martin, P.: Trade-offs and biological diversity: integrative answers to ecological questions. In: Martin, L., Ghalambor, C., Woods, H. (eds.) Integrative Organismal Biology, pp. 291–308. Wiley (2014)
7. Carré, M., Bondarenko, M., Montero, M-P., Chapuisat, G., Benabdallah, A., Le Grand, M., Braguer, D., André, N., Pasquier, E.: Metronomic scheduling: a promising strategy to manage intratumor heterogeneity and control treatment resistance. In: Proceedings of the 106th Annual Meeting of the American Association for Cancer Research, April 18–22 2015, Philadelphia, PA (Cancer Res 75(15 Suppl), 2572 (2015))
8. Warburg, O.: On the origin of cancer cells. Science **123**, 309–314 (1956)
9. Kallinowski, F., Vaupel, P., Runkel, S., Berg, G., Fortmeyer, H.P., Baessler, K.H., Wagner, K., Mueller-Klieser, W., Walenta, S.: Glucose uptake, lactate release, ketone body turnover, metabolic micromilieu, and pH distributions in human breast cancer xenografts in nude rats. Cancer Res. **48**, 7264–7272 (1988)

10. Peppicelli, S., Bianchini, F., Calorini, L.: Extracellular acidity, a "reappreciated" trait of tumor environment driving malignancy: perspectives in diagnosis and therapy. Cancer Metastasis Rev. **33**, 823–832 (2014)
11. Provent, P., Benito, M., Hiba, B., Farion, R., López-Larrubia, P., Ballesteros, P., Rémy, C., Segebarth, C., Cerdán, S., Coles, J., García-Martín, M.: Serial in vivo spectroscopic nuclear magnetic resonance imaging of lactate and extracellular pH in rat gliomas shows redistribution of protons away from sites of glycolysis. Cancer Res. **67**, 7638–7645 (2007)
12. Carmona-Fontaine, C., Bucci, V., Akkari, L., Deforet, M., Joyce, J.A., Xavier, J.B.: Emergence of spatial structure in the tumor microenvironment due to the Warburg effect. Proc. Nat. Acad. Sci. USA **110**, 19402–19407 (2013)
13. Robertson-Tessi, M., Gillies, R.J., Gatenby, R.A., Anderson, A.R.: Impact of metabolic heterogeneity on tumor growth, invasion, and treatment outcomes. Cancer Res. **75**, 1567–1579 (2015)
14. Carmona-Fontaine, C., Deforet, M., Akkari, L., Thompson, C.B., Joyce, J.A., Xavier, J.B.: Metabolic origins of spatial organization in the tumor microenvironment. Proc. Nat. Acad. Sci. USA **114**, 2934–2939 (2017)
15. Hanahan, D., Weinberg, R.: Hallmarks of cancer: the next generation. Cell **144**, 646–674 (2011)
16. Gatenby, R., Silva, A., Gillies, R., Frieden, B.: Adaptive therapy. Cancer Res. **69**, 4894–4903 (2009)
17. Broxterman, H., Pinedo, H., Kuiper, C., Kaptein, L., Schuurhuis, G., Lankelma, J.: Induction by verapamil of a rapid increase in ATP consumption in multidrug-resistant tumor cells. FASEB J. **2**, 2278–2282 (1988)
18. Harada, D., Takigawa, N., Ochi, N., Ninomiya, T., Yasugi, M., Kubo, T., Takeda, H., Ichihara, E., Ohashi, K., Takata, S., Tanimoto, M., Kiura, K.: JAK2-related pathway induces acquired erlotinib resistance in lung cancer cells harboring an epidermal growth factor receptor-activating mutation. Cancer Sci. **103**, 1795–1802 (2012)
19. Wang, Q., Cui, K., Espin-Garcia, O., Cheng, D., Qiu, X., Chen, Z., Moore, M., Bristow, R., Xu, W., Der, S., Liu, G.: Resistance to bleomycin in cancer cell lines is characterized by prolonged doubling time, reduced DNA damage and evasion of G2/M arrest and apoptosis. PLoS ONE **8**, e82363 (2013)
20. Abercrombie, M.: Contact inhibition and malignancy. Nature **281**, 259–262 (1979)
21. Sherratt, J.: Wavefront propagation in a competition equation with a new motility term modelling contact inhibition between cell populations. Proc. R. Soc. A. **456**, 2365–2386 (2000)
22. Sherratt, J., Chaplain, M.: A new mathematical model for avascular tumour growth. J. Math. Biol. **43**, 291–312 (2001)
23. Bertsch, M., Hilhorst, D., Izuhara, H., Mimura, M.: A nonlinear parabolic-hyperbolic system for contact inhibition of cell-growth. Diff. Equ. Appl. **4**, 137–157 (2012)
24. Bertsch, M., Hilhorst, D., Izuhara, H., Mimura, M., Wakasa, T.: Travelling wave solutions of a parabolic-hyperbolic system for contact inhibition of cell-growth. Eur. J. Appl. Math. **26**, 297–323 (2015)
25. Bertsch, M., Passo, R.D., Mimura, M.: A free boundary problem arising in a simplified tumour growth model of contact inhibition. Interfaces Free Bound. **12**, 235–250 (2010)
26. Murray, J.D.: Mathematical Biology. Springer, New York (1993)
27. Maezawa, H., Wong, K., Urano, M.: Radiosensitivity of mouse skin epithelial cell line established in serum-free culture: an alternative to animal use. Int. J. Radiat. Oncol. Biol. Phys. **24**, 533–541 (1992)
28. Gatenby, R., Gawlinski, E.: A reaction-diffusion model of cancer invasion. Cancer Res. **56**, 5745–5733 (1996)
29. Kolmogorov, A., Petrovskii, I., Piscounov, N.: A study of the diffusion equation with increase in the amount of substance, and its application to a biological problem. Bull. Moscow Univ. Math. Mech. **1**, 1–25 (1937)
30. Benga, G.: Basic studies on gene therapy of human malignant melanoma by use of the human interferon β-gene entrapped in cationic multilamellar liposomes.: 1. morphology and growth

rate of six melanoma cell lines used in transfection experiments with the human interferon β-gene. J. Cell. Mol. Med. **5**, 402–408 (2001)
31. Walenta, S., Wetterling, M., Lehrke, M., Schwickert, G., Sundfør, K., Rofstad, E., Mueller-Klieser, W.: High lactate levels predict likelihood of metastases, tumor recurrence, and restricted patient survival in human cervical cancers. Cancer Res. **60**, 916–921 (2000)
32. Foo, J., Michor, F.: Evolution of acquired resistance to anti-cancer therapy. J. Theor. Biol. **355**, 10–20 (2014)
33. Pisco, A.O., Huang, S.: Non-genetic cancer cell plasticity and therapy-induced stemness in tumour relapse: what does not kill me strengthens me. Br. J. Cancer **112**, 1725–1732 (2015)
34. Weiner, R., Schmitt, B., Podhaisky, H.: ROWMAP—a ROW-code with Krylov techniques for large stiff ODEs. Appl. Numer. Math. **25**, 303–319 (1997)
35. Hecht, F.: New development in FreeFem++. J. Numer. Math. **20**, 251–265 (2012)

Efficient Error Based Metrics for Fuzzy-Neural Network Performance Evaluation

Margarita Terziyska, Yancho Todorov and Maria Dobreva

Abstract In this paper the effectiveness of different error metrics for assessment of the capabilities of an advanced fuzzy-neural architecture are studied. The proposed structure combines the potentials of the Intuitionistic Fuzzy Logic with the simplicity of the Neo-Fuzzy Neuron theory for implementation of robust modeling mechanisms, able to capture uncertain variations in the data space. A major concern when evaluating the performance of such kind of models is the selection of appropriate error metrics in order to assess their potential to capture a wide range of system behaviours. Therefore, different error metrics to evaluate the functional properties of a proposed Intuitionistic Neo-fuzzy network are studied and a comparative analysis in modeling of chaotic time series is made.

1 Introduction

Predictive modeling is an essential approach in various application fields where signal and data processing are used to assess interesting data features needed for many up to date engineering tasks. The classification algorithms, the regression models and the factor analysis are well-known as predictive modelling techniques. Recently, such algorithms have been transformed in the framework of machine learning, where the potentials of fuzzy logic and neural networks led to more simple and transparent implementations. Both techniques have a proven advantage over the traditional statistical estimation due to their possibility to deal with uncertain, imprecise and noisy

M. Terziyska · M. Dobreva
Department of Informatics and Statistics, University of Food Technologies,
26 Maritza blvd., 4000 Plovdiv, Bulgaria
e-mail: m.terziyska@uft-plovdiv.bg

M. Dobreva
e-mail: mimi.d.d@abv.bg

Y. Todorov (✉)
Department of Chemical and Metallurgical Engineering, Laboratory of Automation
and Process Control, Aalto University, Kemistintie 1, 02150 Espoo, Finland
e-mail: yancho.todorov@aalto.fi

K. Georgiev et al. (eds.), *Advanced Computing in Industrial Mathematics*, Studies in Computational Intelligence 728,
https://doi.org/10.1007/978-3-319-65530-7_17

data streams. They can estimate or classify any function without the need of precise mathematical description between the input and the target data. A special class within this models are the fuzzy-neural structures. The fusion of the fuzzy logic with the neural networks allow to combine the learning and computational ability of neural networks with the human like IF-THEN thinking and reasoning of a fuzzy system.

A lot of architectures have been proposed in the literature that combine fuzzy logic and neural network. Some of the most popular based on "if-then" notations are described in [12–14]. In principle, the number of fuzzy rules depends exponentially on the number of inputs and membership functions. It derives their main drawback—the huge number of parameters that need to be updated on-line at each sampling period. Another disadvantage of the classical neuro-fuzzy systems, especially when they operate in on-line mode is the slow convergence of the conventional gradient-based learning procedures and the computational complexity of second-order ones. As well, such classical structures cannot handle major process uncertainties in many complex situations.

The idea for a Neo-Fuzzy Neuron (NFN) has been introduced in the early 90s by the works of Uchino and Yamakawa as potential simpler approach for modelling of highly nonlinear systems. The most important properties of the NFNs are their computational simplicity, the proven high approximation properties and the possibility of finding the global minimum of the learning criterion in real time [25, 26].

During the last years, different applications of the NFN concept are reported in the literature. For instance, in [4, 5] the authors propose an approach for on-line linear system parameter estimation using a NFN algorithm and universal approximator employing NFNs. In [6–8] different NFN topologies are presented, while in [5, 16] respective learning approaches are reported. Practical applications to flux observation in induction motors, bearing condition prediction, stock exchange forecasting and bacteria foraging optimization are shown in [15, 18, 22, 29]. An approach to classification task is discussed in [20], as well as an evolving NFN structures are proposed in [21].

In order to be able to handle uncertainties, fuzzy-neural networks are usually equipped with Type-2 fuzzy sets instead of Type-1. Type-2 fuzzy logic was proposed by Zadeh [28] in response to continuing criticism that Type-1 fuzzy sets cant deal with uncertaint data variations. A Type-2 fuzzy set is characterized by a fuzzy membership function, unlike a Type-1 set where the membership grade is a crisp number in [0, 1]. The idea behind Type-2 fuzzy logic is well accepted by the academics and nowadays a lot of scientific papers dedicated to Type-2 Fuzzy Neural Networks can be identified [1, 2, 13, 23, 24]. In the early 80s Atanasov [3] presented its Intuitionistic fuzzy sets theory which is alternative of Type-2 fuzzy logic, as a tool to deal with the vagueness in date space by introducing an additional arbitrary parameter [9].

On the other hand, many fuzzy-neural networks are proposed in the literature, but often is hard to assess their functional properties due to the great number of error metrics used for any particular case. Along with the well-known Root Squared Error (RSE) and Root Mean Squared Error (RMSE), the Mean Absolute Error (MAE), the Mean Absolute Percentage Error (MAPE), the Root Mean Square Percentage Error (RMSPE) and the Mean Relative Absolute Error (MRAE) metrics are rarely discussed. A comprehensive survey of the most common used forecast error measurements is presented in [19], where the authors proposed an Integral Normalized Mean Square Error as a mean to reduce the impact of outliers. Is RMSE or MAE is the better error metric? There is no clear answer to this question, as it depends mostly on the studied case. Usually, it is suggested that the RMSE is not a good indicator of the average model performance and might be a misleading of the average error variations, such that the MAE should be employed [27]. Arguments against avoiding RMSE in the literature are discussed in [10]. Different percentage error metrics are also commented in [11, 17].

Although, many papers discuss different errors such as metrics for evaluating the models, there are no studies which are suitable for neural-fuzzy models. In this paper an Intuitionistic Neo-fuzzy network is proposed and three groups of errors (absolute, relative and percentage) are studied in order to determine their effectiveness in model evaluation. To investigate the functional properties of the proposed network, numerical experiments in modelling of Mackey-Glass chaotic time series are made.

2 Intuitionistic Neo-fuzzy Network

The structure of a Neo-fuzzy neuron is shown in Fig. 1. The Neo-fuzzy neuron functionality is similar to a zero order Sugeno fuzzy system where only one input is included in each fuzzy rule, and to a radial basis function network (RBFN) with scalar arguments of the basis functions [11].

The Neo-fuzzy neuron has a nonlinear synaptic transfer characteristic realized as a set of fuzzy implication rules [26]. Therefore, the output of the neuron is obtained by the following equation:

$$f(x) = \sum_{j=1}^{m} \mu_j(x(k))w_j \tag{1}$$

where $x(k)$ is the input, w_j is the weight coefficient and j for $j = 1 : m$ is a defined set of Gaussian membership functions:

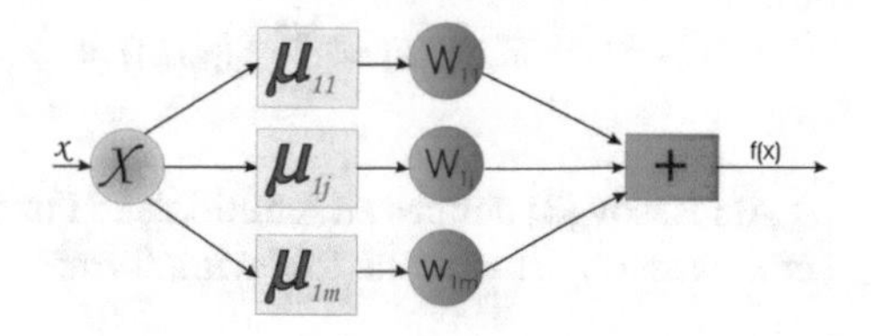

Fig. 1 Structure of a single Neo-fuzzy neuron

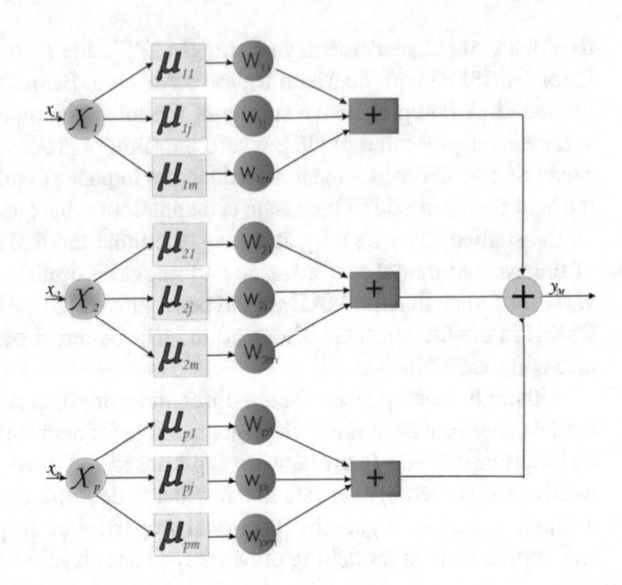

Fig. 2 Structure of Neo-fuzzy network

$$\mu_{Xp,m}^{(n)} = \exp \frac{-(x_p - c_{Xp,m})^2}{2\sigma_{Xp,m}^2} \tag{2}$$

and c represent its centres, while σ (mean) defines their width (standard deviation). Each nonlinear synapse is expressed by a fuzzy rule matching to singleton rule consequents:

$$\text{If } x_j \text{ is } A_{ij} \text{ then the output is } f(x_j) \tag{3}$$

Using the basic concept for a Neo-fuzzy neuron it can be easily designed a network of such neurons capturing the dynamics of a set of multiple inputs and outputs. The typical NFN structure is presented in Fig. 2, where the system output is expressed as:

$$y_m(k) = f(x(k)) \tag{4}$$

and $x(k)$ is an input vector of the states in terms of different time instants. Thus, the output is determined by the membership functions μ_{ji} and the weight coefficients $w_{ji}(k)$:

$$y_m(k) = \sum_{i=1}^{p} f_i(x_i(k)) = \sum_{i=1}^{p} \sum_{j=1}^{m} \mu_{ij}(x_i(k)) w_{ij}(k) \tag{5}$$

Atanassov [3] defines an Intuitionistic Fuzzy Set (IFS) A in over a finite universal set E as an object with the following form:

$$A = \{(x, \mu_A(x), \nu_A(x)) | x \in X\} \tag{6}$$

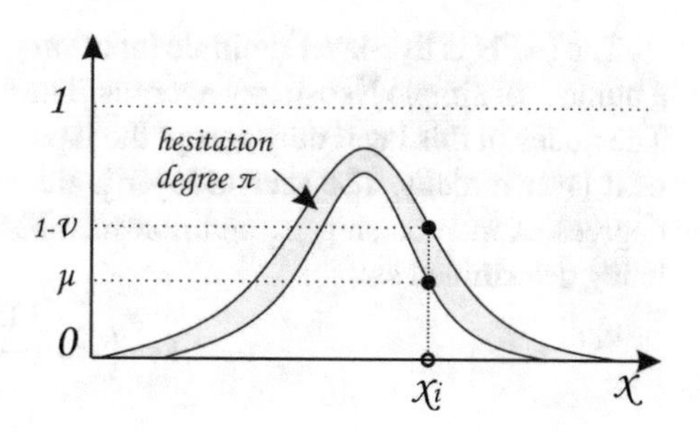

Fig. 3 Representation of an intuitionistic fuzzy set

where $\mu_A : X[0,1]$ and $\nu_A : X[0,1]$ are such that $0 \leq \mu_A + \nu_A \leq 1, \mu_A$ denote a degree of membership of $x \in A$, $\nu_A(x)$ denote a degree of non-membership of $x \in A$. For each intuitionistic fuzzy set in X, we call $\pi_A(x) = 1 - \mu_A - \nu_A$ the degree on non-determinacy (uncertainty) or hesitation of $x \in A$. This parameter expresses the hesitation degree of whether $x \in A$ or not and it is obviously $0 \leq \pi_A \leq 1$ for each $x \in X$. On Fig. 3 is shown the representation of an Intuitionistic fuzzy set.

Combining the advantages of the both AI paradigms, an Intuitionistic neo-fuzzy network (INFN) can be designed as a simpler structure able to operate with uncertain data variations in a computationally effective manner. The structure of the proposed network is shown on Fig. 4.

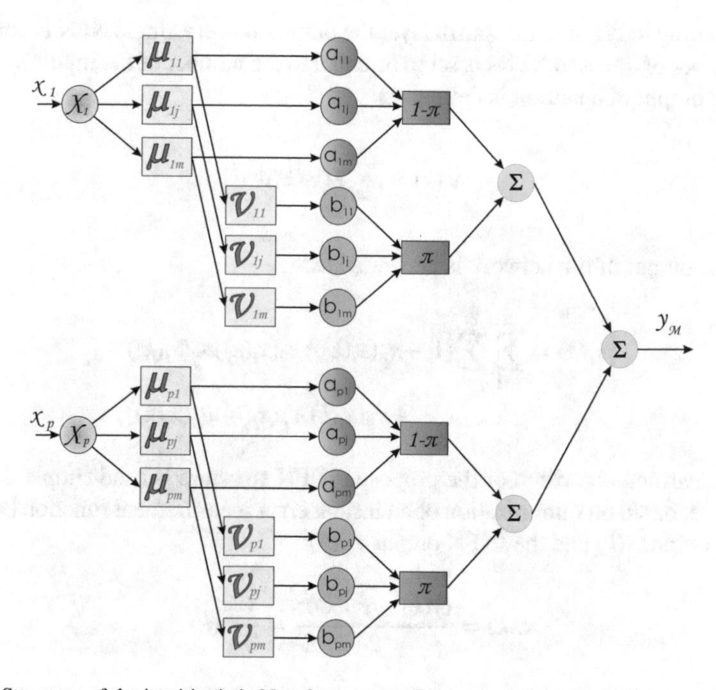

Fig. 4 Structure of the intuitionistic Neo-fuzzy network

The INFN is five-layer multiple input single output (MISO) structure that consists a number of simple Neo-fuzzy neurons. The first layer of the INFN is the input layer. The nodes in this layer only accept the input variables and then transmit them to the next layer directly. The second layer is the so-called fuzzification layer, where the degrees of membership μ_A and non-membership ν_A, using Gaussian functions are being determined as:

$$\mu_{ij}(x_i) = \left(exp\frac{-(x_p - c_i)^2}{2\sigma_i^2} \right) \tag{7}$$

$$\nu_{ij}(x_i) = \left(1 - exp\left(\frac{-(x_p - c_i)^2}{2\sigma_i^2} \right) \right)^k, k \geq 1 \tag{8}$$

where x_i is the input value, $i = 1 : p$ is the number of the inputs of the IFNF, c_{ij} and σ_{ij} are the center and the standard deviation of the Gaussian membership function, $j = 1 : m$ where m is the number of used membership functions, k is parameter that must be designed. If the $k = 0$, then obviously $\mu_A + \nu_A = 1$ and the hesitation degree π_A (which is also computed on this layer) also is zero. The neurons of the third layer calculate the following expression:

$$f_i(x_i(k)) = (1 - \pi_{ij}(x_i(k)))\mu_{ij}(x_i(k))a_{ij}(x_i(k)) + \pi_{ij}(x_i(k))\nu_{ij}(x_i(k))b_{ij}(x_i(k)) \tag{9}$$

According to Fig. 4 in the fourth layer the output of every single NFN is obtained. The number of the used NFNs is set to be equal to the number of the inputs p. Therefore, the output of a neuron is defined as:

$$y_i(k) = \sum_{i=1}^{m} f_i(x_i(k)) \tag{10}$$

while the output of the network is generated as:

$$y_M(k) = \sum_{i=1}^{p} \sum_{j=1}^{m} (1 - \pi_{ij}(x_i(k)))\mu_{ij}(x_i(k))a_{ij}(x_i(k)) \\ + \pi_{ij}(x_i(k))\nu_{ij}(x_i(k))b_{ij}(x_i(k)) \tag{11}$$

As a learning algorithm of the proposed INFN structure is used simple gradient procedure, based on minimization of an instant error measurement function between the real output $y(k)$ and the INFN output y_M(k):

$$E(k) = \frac{(y(k) - y_M(k))^2}{2} = \frac{1}{2}e^2(k) \tag{12}$$

During the learning procedure only the a_{ji} and b_{ij} parameters are adjusted while the membership μ_A and non-membership ν_A functions are not trained. The adaptation of the membership and non-membership functions will be discussed in our future works. From (11) it could be easily derived the updating rules:

$$a(k+1) = a(k) + \eta\left(\frac{\partial E(k)}{\partial a(k)}\right) \tag{13}$$

$$b(k+1) = b(k) + \eta\left(\frac{\partial E(k)}{\partial b(k)}\right) \tag{14}$$

where η is the learning rate, a and b are the vectors of the trained parameters (the synaptic links in the consequent part of the zero order Sugeno rules associated with the membership and non-membership functions respectively). Finally, after calculating the partial derivatives in (13) and (14), the recurrent learning rules can be expressed as:

$$a(k+1) = a(k) + \eta e(k)(1 - \pi(k))\mu(k) \tag{15}$$

$$b(k+1) = b(k) + \eta e(k)\pi(k)\nu(k) \tag{16}$$

3 Metrics for Evaluation of the Intuitionistic Neo-fuzzy Network Performance

The proposed Intuitionistic Neo-fuzzy network performances were evaluated using coefficient of determination (R^2) and three groups of errors—absolute errors, relative errors and percentage errors.

3.1 *Coefficient of Determination*

A classical way to summarize how well a model fits the data is via the coefficient of determination or R^2. If the estimations are close to the actual values, we would expect R^2 to be closer to 1. On the other hand, if the estimations are unrelated to the actual values, then $R^2 = 0$. In all cases, R^2 lies between 0 and 1. The coefficient of determination can be calculated as the square of the correlation between the observed y values and the estimated $\hat{y}$ values:

$$R^2 = \left(1 - \frac{SSE}{SS_{yy}}\right) \tag{17}$$

where

$$SSE = \sum_{i=1}^{n}(y_i - \hat{y}_i)^2 \tag{18}$$

$$SS_{yy} = \sum_{i=1}^{n} (y_i - \bar{y})^2 = \sum_{i=1}^{n} y_i^2 - \frac{\left(\sum_{i=1}^{n} y_i\right)^2}{n} \tag{19}$$

Alternatively, it can be calculated also as:

$$R^2 = \frac{\sum(\hat{y}_i - \bar{y})^2}{\sum(y_i - \bar{y})^2} \tag{20}$$

where the summations are over all observations. Thus, it is also the proportion of variation in the forecast variable that is accounted for (or explained) by the Intuitionistic Neo-fuzzy model.

3.2 Absolute Errors

This group of errors is based on the absolute error calculation. It includes estimates based on the calculation of the value e_i:

$$e_i = (y_i - \hat{y}_i) \tag{21}$$

The most popular absolute errors are Mean Absolute Error (MAE), Mean Square Error (MSE) and Root Mean Square Error (RMSE):

$$MAE = \frac{1}{n} \sum_{i=1}^{n} |e_i| \tag{22}$$

$$MSE = \frac{1}{n} \sum_{i=1}^{n} \left(e_i^2\right) \tag{23}$$

$$RMSE = \sqrt{\frac{1}{n} \sum_{i=1}^{n} (e_i^2)} \tag{24}$$

These errors are widely used in various domains, because they are simple and easy to compute. MAE and RMSE are scale-dependent measures which is their main drawback. MAE has the same units as the original data, and it can only be compared between models whose errors are measured in the same units. Usually, it is similar in magnitude to RMSE, but slightly smaller. The high influence of outliers is another disadvantage of the absolute errors. MAE is more robust to outliers than is MSE. MAE assigns equal weight to the data whereas MSE emphasizes the extremes—the square of a very small number is even smaller, and the square of a big number is even bigger. RMSE basically tells you to avoid models that give you occasional large errors. RMSE can only be compared between models whose errors are measured in the same units. Mean Squared Error (MSE) is the most common measure of numerical model performance. It simply represents the average of the squares of the differences between the predicted and actual values. It is a reasonably good measure of performance, though it could be argued that it overemphasizes the importance of larger errors. Many modeling procedures directly minimize the MSE.

3.3 *Relative Errors*

Relative errors are another valuable metric of performance evaluation of the proposed Intuitionistic Neo-fuzzy model. The basis for calculation of errors in this group is the relative error, which is the absolute error divided by the actual measurement. Since, the actual measurement is unknown, the measured value is used. In this paper from the group of relative errors, the Relative Squared Error (RSE) and the Root Relative Squared Error (RRSE) are considered. RSE can be compared between models whose errors are measured in the different units. It can be computed easily using the following equation:

$$RSE = \frac{\sum_{i=1}^{n}(\hat{y}_i - y_i)^2}{\sum_{i=1}^{n}(y_i - \bar{y})^2} \tag{25}$$

RRSE has a scale from 0 to 1 and when is multiplied by 100 it gets similarly in 0–100 scale (i.e. percentage). Therefore, smaller values are preferred.

$$RRSE = \sqrt{\frac{\sum_{i=1}^{n}(\hat{y}_i - y_i)^2}{\sum_{i=1}^{2}(y_i - \bar{y})^2}} \tag{26}$$

3.4 *Percentage Errors*

The third group of error metrics for model evaluation is the group of percentage errors. The percentage error is the relative error shown as a percentage (27). This kind of errors have the advantage of being scale-independent, and so they are frequently used to compare forecast performance between different data sets.

$$p_i = \frac{100e_i}{y_i} \tag{27}$$

Measures based on the percentage errors have the disadvantage of being infinite or undefined if $y_i = 0$ for any i in the period of interest, and having extreme values when any y_i is close to zero. Another problem with percentage errors that outliers have significant impact on the result and that the error measures are biased. The most popular percentage errors are Mean Percentage Error (MPE), Mean Absolute Percentage Error (MAPE) and Root Mean Square Percentage Error (RMSPE):

$$MPE = \frac{100}{n}\sum_{i=1}^{n}\frac{y_i - \hat{y}_i}{y_i} \tag{28}$$

$$MAPE = \frac{1}{n}\sum_{i=1}^{n}\frac{|e_i|}{y_i} * 100 \tag{29}$$

$$RMSPE = \sqrt{\left(\frac{1}{n}\sum_{i=1}^{n}\frac{|e_i|}{y_i} * 100\right)^2} \tag{30}$$

4 Numerical Experiments

To investigate the modelling potentials of the proposed Intuitionistic Neo-fuzzy model, the Mackey-Glass (MG) chaotic times series benchmark have been used. The MG time series is described by the following time-delay differential equation:

$$x(i+1) = \frac{x(i) + ax(i-s)}{(1 + x^c(i-s)) - bx(i)} \tag{31}$$

where $a = 0.2$; $b = 0.1$; $C = 10$; initial conditions $x(0) = 0.1$ and $s = 17$ s. Results on Intuitionistic Neo-fuzzy model validation by using Mackey-Glass chaotic time series are shown in Fig. 5. The evolution of adopted error metrics in time in their natural scale—in Figs. 6, 8 and 10, and in logarithmic scale—Figs. 7, 9 and 11 is presented. The values of the studied metrics are show in Tables 1, 2 and 3. According to (17)–(19), the coefficient of determination is calculated. In this particular case it has a value of $R^2 = 94, 15\%$, that proves the good generalization properties of the studied modeling structure.

The conducted experiments show that the proposed modeling structure perform well in estimating the oscillating times series data, and the error between the reference and the modeled outputs is successfully minimized. The adopted logarithmic scale show in a more clearer manner the tendency of decrease, as well as the similar dynamical nature of the studied error terms. The achieved results are not surprising, since a simple instant error measurement is taken as reference point for calculation of all of the proposed metrics. Therefore, the specifics of every metric is focused mostly to assess the impact caused by certain data phenomena, e.g. outliers. On the other hand, it should be mentioned that choosing an appropriate error metrics will strongly depend on the purpose of the fuzzy-neural network. If the modeling is being performed off-line and the data has unknown features, the application of the three different groups of metrics will give a valuable information on the impact of those

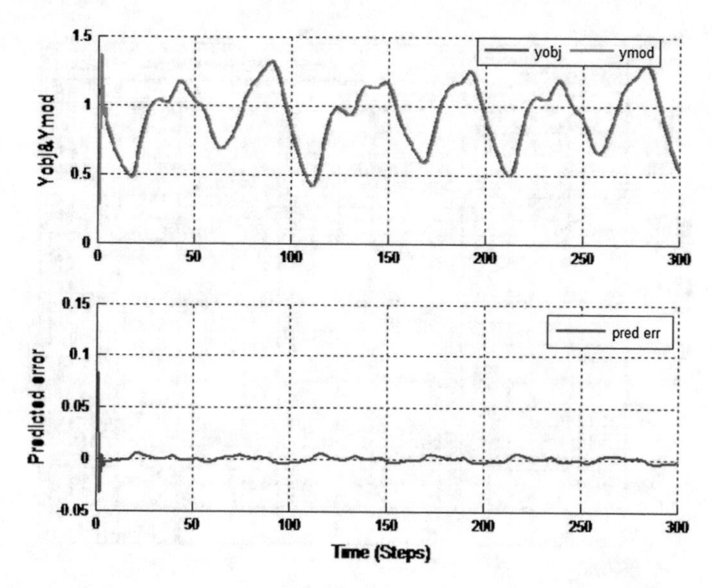

Fig. 5 Intuitionistic Neo-fuzzy model validation by using Mackey-Glass chaotic time series

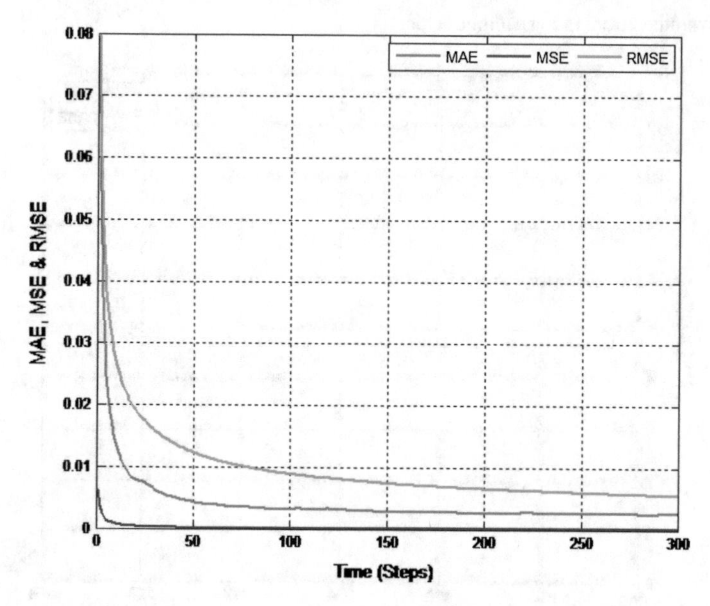

Fig. 6 Absolute errors

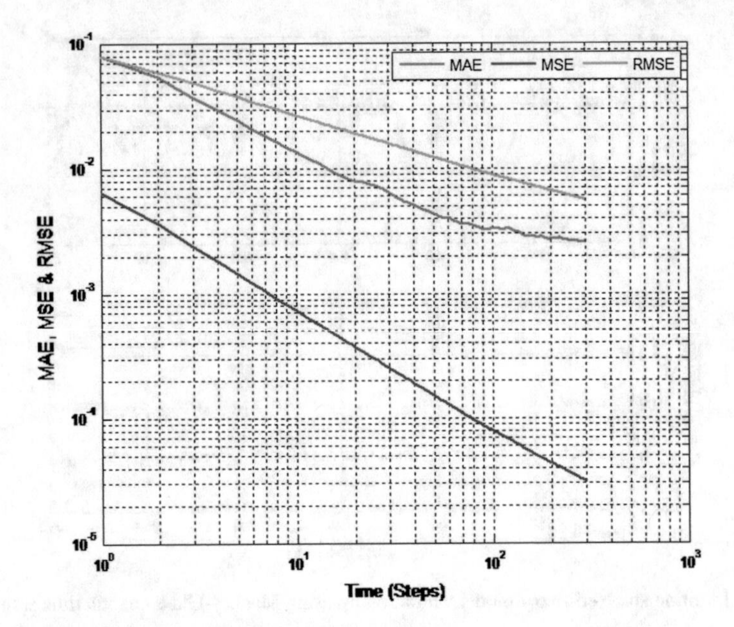

Fig. 7 Absolute errors in logarithmic scale

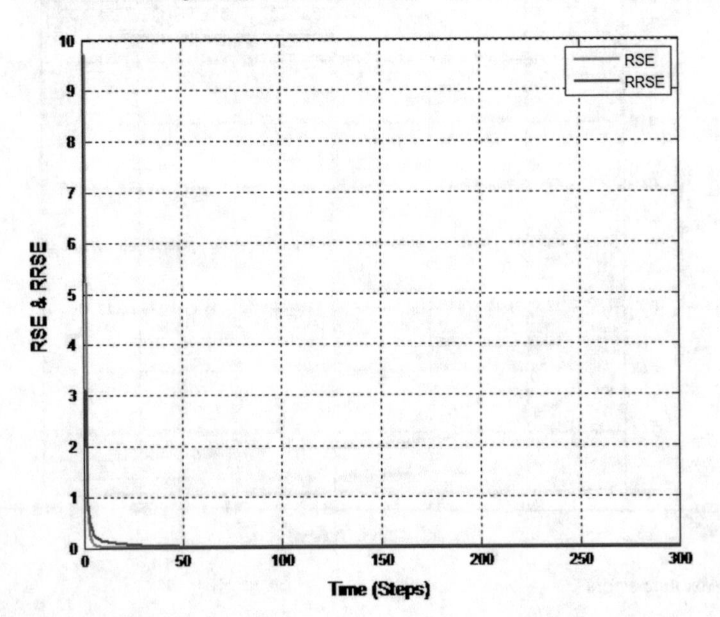

Fig. 8 Relative errors

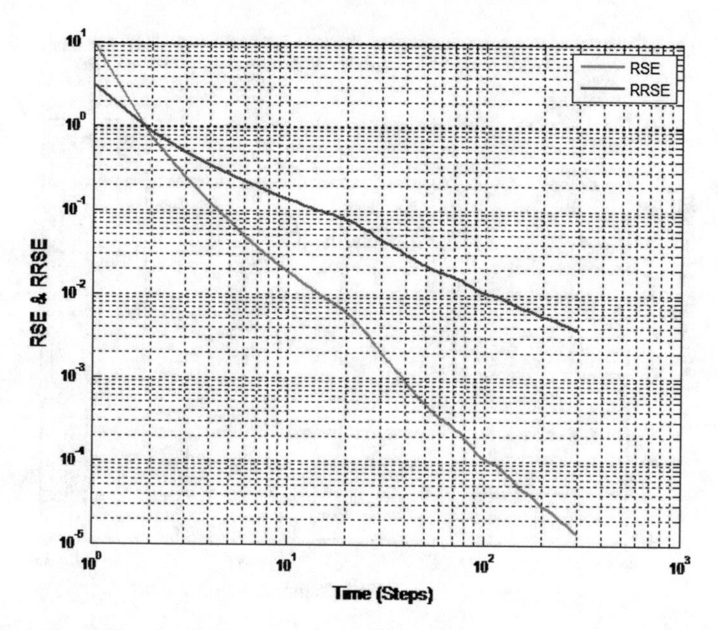

Fig. 9 Relative errors in logarithmic scale

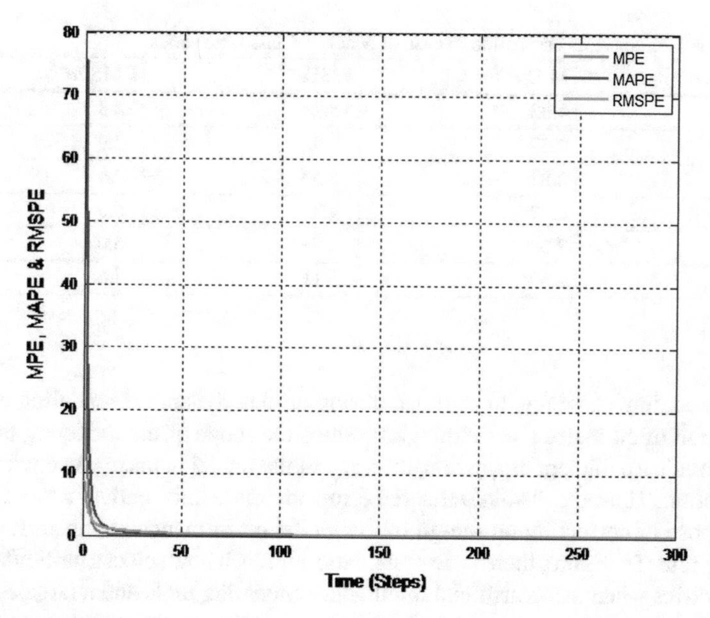

Fig. 10 Percentage errors

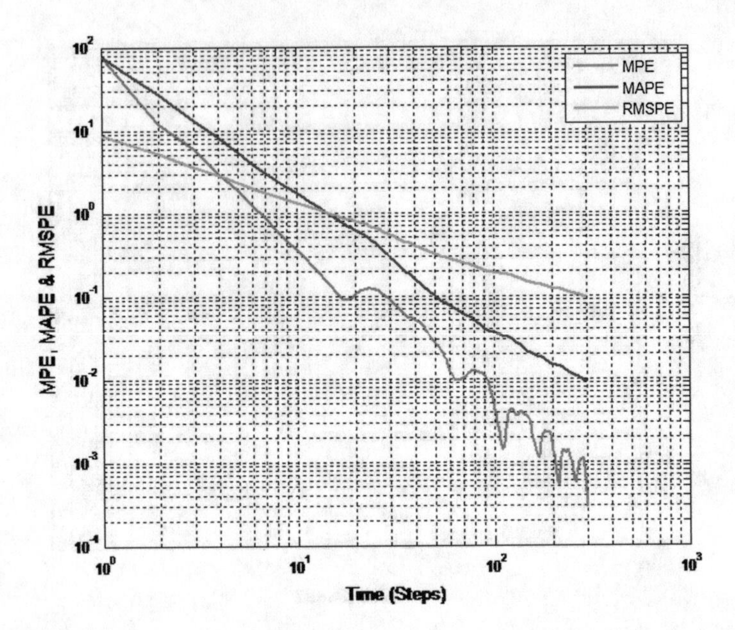

Fig. 11 Percentage errors in logarithmic scale

Table 1 Absolute errors estimation in case of Mackey-Glass time series

Time steps	MAEe^{-2}	MSEe^{-5}	RMSEe^{-5}
50	4.33	15.2	12.3
100	3.27	7.92	8.9
150	2.83	5.50	7.4
200	2.72	4.31	6.6
250	2.60	3.58	6.0
300	2.59	3.11	5.6

on the modeling accuracy. In contrast, during on-line dynamical modeling case the application of all metrics is meaningless, since the scope of the modeling task will be to provide a sufficient estimation accuracy of the model in uncertain environment of operation. Thus, the absolute squared errors are more informative in this case, for the purpose of correcting on-the-go of the model performance parameters, e.g. the learning rate. Therefore, there is no systematic approach how select a particular set of error metrics when using artificial intelligence modeling tools and a targeted selection of such should be performed beforehand, depending on the concrete application area.

Table 2 Relative errors estimation in case of Mackey-Glass time series

Time steps	RSEe^{-3}	RRSEe^{-3}
50	641	253
100	0.428	2.1
150	0.0946	9.7
200	0.0452	6.7
250	0.0259	5.1
300	0.0173	4.2

Table 3 Percentage errors estimation in case of Mackey-Glass time series

Time steps	MPE %	MAPE %	RMSPE %
0	71.680	71.680	8.471
50	0.026	0.098	0.312
100	0.0045	0.035	0.187
150	0.0033	0.021	0.143
200	0.0013	0.015	0.121
250	8.54e^{-4}	0.011	0.106

5 Conclusions

It was presented in this paper a comparative study on the properties of different error metrics when modeling a typical fuzzy-neural network. The scope of the study was tailored to assess different measures deriving from the three of the most commonly used groups: the absolute, relative and percentage errors. The performed experiments and the collected results shown that, no matter of the chosen metric the studied error terms have a similar nature and tendency to decrease, where the typical values depend on the units of the modeled system. If an uncertain variation of the data space is experienced, that is expected to lead of increased variations in all of the studied metrics. On the other hand, most of the recent AI based modeling structures are applied for on-line estimation of data with unknown properties, that persistently stimulates the works on handling uncertain data space variations. In that sense, the application for on-line purposes of relative and percentage errors will be meaningless, while the metrics based on absolute error estimation will be more informative for evaluation of the model performances on the go.

References

1. Abiyev, R.H., Kaynak, O.: Type 2 fuzzy neural structure for identification and control of time-varying plants. IEEE Trans. Ind. Electron. **57**(12) (2010)
2. Aliev, R.A., Guirimov, B.G.: Type-2 Fuzzy Neural Networks and Their Applications. Springer International Publishing (2014)
3. Atanassov, K.: Intuitionistic Fuzzy Sets. Springr, Hielderberg (1999)
4. Baceiar, A., De Souza Filho, E., Neves, F., Landim, R.: On-line linear system parameter estimation using the neo-fuzzy-neuron algorithm. In: Proceedings of the 2nd IEEE International Workshop on Intelligent Data Acquisition and Advanced Computing Systems: Technology and Applications, pp. 115–118 (2003)
5. Bodyanskiy, Y., Kokshenevane, I., Kolodyazhniy, V.: An adaptive learning algorithm for a neo-fuzzy neuron. In: Proceedings of the 3rd Conference of the European Society for Fuzzy Logic and Technology, pp. 375–379 (2005)
6. Bodyanskiy, Y., Viktorov, Y.: The cascade neo-fuzzy architecture and its online learning algorithm. Int. Book Series Inf. Sci. Comput. **17**(1), 110–116 (2010)
7. Bodyanskiy, Y., Pliss, I., Vynokurova, O.: Flexible neo-fuzzy neuron and neuro-fuzzy network for monitoring time series properties. Inf. Technol. Manag. Sci. **16**, 47–52 (2013)
8. Bodyanskiy, Y., Tyshchenko, O., Kopaliani, D.: An extended neo-fuzzy neuron and its adaptive learning algorithm. Int. J. Intell. Syst. Appl. **2**, 21–26 (2015)
9. Castillo, O., Melin, P., Tsvetkov, R., Atanassov, K.T.: Short remark on fuzzy sets, interval type-2 fuzzy sets, general type-2 fuzzy sets and intuitionistic fuzzy sets. In: Intelligent Systems' 2014, Advances in Intelligent Systems and Computing, vol. 322, pp. 183–190 (2015)
10. Chai, T., Draxler, R.R.: Root mean square error (RMSE) or mean absolute error (MAE)?–Arguments against avoiding RMSE in the literature. Geosci. Model Dev. **7**, 1247–1250 (2014)
11. Collopy, F., Armstrong, J.S.: Another Error Measure for Selection of the Best Forecasting Method: The Unbiased Absolute Percentage Error (2000)
12. Jang, J.-S.R.: ANFIS: adaptive-network-based fuzzy inference system. IEEE Trans. Syst. Man Cybern. **23**(5), 665–685 (1993)
13. Juang, C.-F., Jang, W.-S.: A type-2 neural fuzzy system learned through type-1 fuzzy rules andits FPGA-based hardware implementation. Appl. Soft Comput. **18**, 302–313 (2014)
14. Kasabov, N.K., Song, Q.: DENFIS: dynamic evolving neural-fuzzy inference system and its application for time-series prediction. IEEE Trans. Fuzzy Syst. **10**(2), (2002)
15. Kim, H.D.: Optimal learning of neo-fuzzy structure using bacteria foraging optimization. In: Proceedings of the ICCA (2005)
16. Landim, R., Rodriguez, B., Silva, S., Caminhas, W.: A neo-fuzzy-neuron with real time training applied to flux observer for an induction motor. In: Proceedings of 5th IEEE Brazilian Symposium of Neural Networks, pp. 67–72 (1998)
17. Moreno, J.J.M., Pol, A.P., Abad, A.S., Blasco, B.C.: Using the RMAPE index as a resistant measure of forecast accuracy. Psicothema **25**(4), 500–506 (2013)
18. Pandit, M., Srivastava, L., Singh, V.: On-line voltage security assessment using modified neo fuzzy neuron based classifier. IEEE Int. Conf. Ind. Technol. 899–904 (2006)
19. Shcherbakov, M., Brebels, A.: A survey of forecast error measures. World Appl. Sci. J. **24**(4), 171–176 (2013)
20. Silva, A.M., Caminhas, W., Lemos, A., Gomide, F.: A fast learning algorithm for evolving neo-fuzzy neuron. Appl. Soft Comput. **14**, 194–209 (2014)
21. Silva, A.M., Caminhas, W., Lemos, A., Gomide, F.: Evolving neo-fuzzy neural network with adaptive feature selection. In: Proceedings of 2013 BRICS IEEE Congress on Computational Intelligence, 11th Brazilian Congress on Computational Intelligence, pp. 341–349 (2014)
22. Soualhi, A., Clerc, G., Razik, H., Rivas, F.: Long-term prediction of bearing condition by the neo-fuzzy neuron. In: Proceedings of 9th International IEEE Symposium of Diagnostic of Electric Machines, Power Electronics and Drives, pp. 586–591 (2013)

23. Terziyska, M., Todorov, Y.: Modeling of chaotic time series by interval type-2 neo-fuzzy neural network. In: International Conference on Artificial Neural Networks (ICANN' 2014). Springer Lecture Notes on Computer Science, vol. 8681, pp. 643–650. Hamburg, Germany (2014)
24. Tung, S.W., Quek, C., Guan, C.: T2FIS: an evolving type-2 neural fuzzy inference system. Inf. Sci. **220**, 124–148 (2013)
25. Uchino, E., Yamakawa, T.: Neo-fuzzy-neuron based new approach to system modeling, with application to actual system. In: Proceedings of 6th International IEEE Conference on Tools with Artificial Intelligence, pp. 564–570 (1994)
26. Uchino, E., Yamakawa, T.: High speed fuzzy learning machine with guarantee of global minimum and its applications to chaotic system identification and medical image processing. In: Proceedings of 7th International IEEE Conference on Tools with Artificial Intelligence, pp. 242–249 (1995)
27. Willmott, C., Matsuura, K.: Advantages of the mean absolute error (MAE) over the root mean square error (RMSE) in assessing average model performance. Clim. Res. **30**, 79–82 (2005)
28. Zadeh, L.A.: The concept of a linguistic variable and its applications to approximate reasoning-1. Inf. Sci. **8**, 199–249 (1975)
29. Zaychenko, Y., Gasanov, A.: Investigations of cascade neo-fuzzy neural networks in the problem of forecasting at the stock exchange. In: Proceedings of the IVth IEEE International Conference Problems of Cybernetics and Informatics (PCI' 2012), pp. 227–229 (2012)

Box Model of Migration in Channels of Migration Networks

Nikolay K. Vitanov, Kaloyan N. Vitanov and Tsvetelina Ivanova

Abstract We discuss a box model of migration in channels of networks with possible application for modelling motion of migrants in migration networks. The channel consists of nodes of the network (nodes may be considered as boxes representing countries) and edges that connect these nodes and represent possible ways for motion of migrants. The nodes of the migration channel have different "leakage", i.e. the probability of change of the status of a migrant (from migrant to non-migrant) may be different in the different countries along the channel. In addition the nodes far from the entry node of the channel may be more attractive for migrants in comparison to the nodes around the entry node of the channel. We discuss below channels containing infinite number of nodes. Two regimes of functioning of these channels are studied: stationary regime and non-stationary regime. In the stationary regime of the functioning of the channel the distribution of migrants in the countries of the channel is described by a distribution that contains as particular case the Waring distribution. In the non-stationary regime of functioning of the channel one observes exponential increase or exponential decrease of the number of migrants in the countries of the channel. It depends on the situation in the entry country of the channel for which scenario will be realized. Despite the non-stationary regime of the functioning of the channel the asymptotic distribution of the migrants in the nodes of the channel is stationary. From the point of view of the characteristic features of the migrants we discuss the cases of (i) migrants having the same characteristics and (ii) two classes of migrants that have differences in some characteristic (e.g., different religions).

N.K. Vitanov (✉)
Institute of Mechanics, Bulgarian Academy of Sciences,
Acad. G. Bonchev Str., Bl. 4, 1113 Sofia, Bulgaria
e-mail: vitanov@imbm.bas.bg

K.N. Vitanov
Faculty of Mathematics and Informatics, "St. Kliment Ohridsky" University of Sofia,
James Bourcher Blvd. 5, 1164 Sofia, Bulgaria

T. Ivanova
Faculty of Physics, "St. Kliment Ohridsky" University of Sofia,
James Bourcher Blvd. 5, 1164 Sofia, Bulgaria

K. Georgiev et al. (eds.), *Advanced Computing in Industrial Mathematics*, Studies in Computational Intelligence 728,
https://doi.org/10.1007/978-3-319-65530-7_18

1 Introduction

Flows in complex networks are important for existence and functioning of the systems containing such networks. Human migration (the permanent or semipermanent change of residence that involves e.g., the relocation of individuals, households or moving groups between geographical locations [16]) is one example of such a flow [8, 10]. Large external migration flows reached Europe in the last years and this makes the study of migration very actual topic. In addition the internal migration studies are important for taking decisions about economic development of regions of a country [1–4, 7, 9, 11, 22, 23]. Examples for results from such studies is e.g., the Heckscher–Ohlin theorem for economic use of a country relative abundant factors such as labor as well as the factor-price equalization theorem [2]. Migrant flows may be modelled by deterministic or stochastic tools [12, 15, 17–19, 32] and the corresponding migration models can be classified as probability models [33, 34] or deterministic models with respect to their mathematical features. Examples for probability models are the exponential model, multinomial model or Markov chain models of migration. One of the most famous deterministic models of migration is the gravity model of migration [9]. The gravity model may be extended in different ways, e.g., to include the income and unemployment in the two regions.

Human migration is closely connected to ideological struggles [27, 28] and waves and statistical distributions in population systems [25, 29–31]. In this article we shall consider a box model of a flow of migrants in a channel (sequence of countries) of a migration network. The nodes (countries) will be considered as boxes (cells) where the following processes happen: inflow and outflow of migrants and "leakage" (change of the status of migrant). Migrants enter the channel from the entry country and move through the channel. The different nodes of the channel (the different countries) are assumed to have different rate of "leakage" (i.e. different probabilities of change of the status of a person form the migrant to non-migrant).

The paper is organized as follows. In Sect. 2 the model for moving of substance in a channel containing an infinite number of nodes is discussed. Two regimes of functioning of the channel: stationary regime (the amount of the substance in the entry box of the channel doesn't change) and non-stationary regime (amount of substance in the entry box of the channel decreases or increases exponentially) are described. Statistical distributions of the amount of substance in the nodes of the channel are obtained. A particular case of the distribution for the stationary regime of functioning of the channel is the Waring distribution. Section 3 is devoted to the case of two immiscible substances moving through the channel. In Sect. 4 we relate the obtained mathematical results to the movement of migrants through a migration channel.

2 Channel Containing Infinite Number of Nodes

Inspired by the models in [20, 24, 26] we consider a model of moving of a substance through a channel as follows. The channel contains infinite number of nodes and each node can be considered as a cell. The cells are indexed in succession by non-negative integers. The first cell has index 0. We assume that an amount x of some substance is distributed among the cells and this substance can move from one cell to another cell. Let x_i be the amount of the substance in the i-th cell. Then

$$x = \sum_{i=0}^{\infty} x_i \tag{1}$$

The fractions $y_i = x_i/x$ can be considered as probability values of distribution of a discrete random variable ζ

$$y_i = p(\zeta = i),\ i = 0, 1, \ldots \tag{2}$$

The content x_i of any cell can change because of the following 3 processes:

1. Some amount s of the substance x enters the channel from the external environment through the 0-th cell;
2. Rate f_i from x_i is transferred from the i-th cell into the $i+1$-th cell;
3. Rate g_i from x_i leaks out the i-th cell into the external environment.

The above processes can be modeled mathematically by the system of ordinary differential equations:

$$\begin{aligned}\frac{dx_0}{dt} &= s - f_0 - g_0;\\ \frac{dx_i}{dt} &= f_{i-1} - f_i - g_i,\ i = 1, 2, \ldots.\end{aligned} \tag{3}$$

The following forms of the amount of the moving substances may be assumed ($\alpha, \beta, \gamma_i, \sigma$ are constants)

$$\begin{aligned}&s = \sigma x_0;\quad \sigma > 0\\ &f_i = (\alpha + \beta i)x_i;\quad \alpha > 0,\ \beta \geq 0 \rightarrow \text{cumulative advantage of higher cells}\\ &g_i = \gamma_i x_i;\quad \gamma_i \geq 0 \rightarrow \text{non-uniform leakage over the cells}\end{aligned} \tag{4}$$

The rules (4) differ from the rules in [20] as follows:

1. s is proportional to the amount of the substance x_0 in the 0-th node. In [20] s is proportional to the amount x of the substance in the entire channel;
2. Leakage rates γ_i are different for the different nodes. In [20, 26] the leakage rate is constant and equal to γ for all nodes of the channel (i.e., there is uniform leakage over the cells).

Substitution of Eq. (4) in Eq. (3) leads to the relationships

$$\frac{dx_0}{dt} = \sigma x_0 - \alpha x_0 - \gamma_0 x_0;$$
$$\frac{dx_i}{dt} = [\alpha + \beta(i-1)]x_{i-1} - (\alpha + \beta i + \gamma_i)x_i; \quad i = 1, 2, \ldots \tag{5}$$

There are two regimes of functioning of the channel and realization of one of them depends on the situation in the 0-th node (the entry cell). The regimes are stationary regime and non-stationary regime.

2.1 Stationary Regime of Functioning of the Channel

In the stationary regime of the functioning of the channel $\sigma = \alpha + \gamma_0$ which means that x_0 (the amount of the substance in the 0-th cell of the channel) is free parameter. In this case the solution of Eq. (5) is

$$x_i = x_i^* + \sum_{j=0}^{i} b_{ij} \exp[-(\alpha + \beta j + \gamma_j)t] \tag{6}$$

where x_i^* is the stationary part of the solution. For x_i^* one obtains the relationship

$$x_i^* = \frac{\alpha + \beta(i-1)}{\alpha + \beta i + \gamma_i} x_{i-1}^* \tag{7}$$

The corresponding relationships for the coefficients b_{ij} are

$$b_{ij} = \frac{\alpha + \beta(i-1)}{\gamma_i - \gamma_j + \beta(i-j)} b_{i-1,j}, \; j = 0, 1, \ldots, i-1 \tag{8}$$

From Eq. (7) one obtains

$$x_i^* = \frac{[k + (i-1)]!}{(k-1)! \prod_{j=1}^{i} (k + j + a_j)} x_0^* \tag{9}$$

where $k = \alpha/\beta$ and $a_j = \gamma_j/\beta$. The form of the corresponding stationary distribution $y_i^* = x_i^*/x^*$ (where x^* is the amount of the substance in all of the cells of the channel) is

$$y_i^* = \frac{[k + (i-1)]!}{(k-1)! \prod_{j=1}^{i} (k + j + a_j)} y_0^* \tag{10}$$

Let us consider the particular case where $a_0 = a_1 = \cdots = a$. In this case the distribution from Eq. (10) is reduced to the distribution:

$$P(\zeta = i) = P(\zeta = 0)\frac{(k-1)^{[i]}}{(a+k)^{[i]}}; \quad k^{[i]} = \frac{(k+i)!}{k!}; \; i = 1, 2, \ldots \tag{11}$$

$P(\zeta = 0) = y_0^* = x_0^*/x^*$ is the percentage of substance that is located in the first cell of the channel. Let this percentage be

$$y_0^* = \frac{a}{a+k} \tag{12}$$

The case described by Eq. (11) corresponds to the situation where the amount of substance in the first cell is proportional of the amount of substance in the entire channel (self-reproduction property of the substance). In this case Eq. (10) is reduced to the distribution:

$$P(\zeta = i) = \frac{a}{a+k}\frac{(k-1)^{[i]}}{(a+k)^{[i]}}; \quad k^{[i]} = \frac{(k+i)!}{k!}; \; i = 1, 2, \ldots \tag{13}$$

Let us denote $\rho = a$ and $k = l$. The distribution (13) is exactly the Waring distribution (probability distribution of non-negative integers named after Edward Waring—a Lucasian professor of Mathematics in Cambridge in the 18th century) [6, 13, 14]

$$p_l = \rho\frac{\alpha_{(l)}}{(\rho+\alpha)_{(l+1)}}; \quad \alpha_{(l)} = \alpha(\alpha+1)\ldots(\alpha+l-1) \tag{14}$$

Waring distribution may be written also as follows

$$\begin{aligned} p_0 &= \rho\frac{\alpha_{(0)}}{(\rho+\alpha)_{(1)}} = \frac{\rho}{\alpha+\rho} \\ p_l &= \frac{\alpha+(l-1)}{\alpha+\rho+l}p_{l-1}. \end{aligned} \tag{15}$$

The mean μ (the expected value) of the Waring distribution is

$$\mu = \frac{\alpha}{\rho-1} \text{ if } \rho > 1 \tag{16}$$

The variance of the Waring distribution is

$$V = \frac{\alpha\rho(\alpha+\rho-1)}{(\rho-1)^2(\rho-2)} \text{ if } \rho > 2 \tag{17}$$

ρ is called the tail parameter as it controls the tail of the Waring distribution. Waring distribution contains various distributions as particular cases. Let $i \to \infty$ Then the Waring distribution is reduced to

$$p_l \approx \frac{1}{l^{(1+\rho)}}. \tag{18}$$

which is the frequency form of the Zipf distribution [5]. If $\alpha \to 0$ the Waring distribution is reduced to the Yule–Simon distribution [21]

$$p(\zeta = l \mid \zeta > 0) = \rho B(\rho + 1, l) \tag{19}$$

where B is the beta-function.

2.2 Non-stationary Regime of Functioning of the Channel

In the nonstationary case $dx_0/dt \neq 0$. In this case the solution of the first equation of the system of equation (5) is

$$x_0 = b_{00} \exp[(\sigma - \alpha - \gamma_0)t] \tag{20}$$

where b_{00} is a constant of integration. x_i must be obtained by solution of the corresponding Eq. (5). The form of x_i is

$$x_i = \sum_{j=0}^{i} b_{ij} \exp[-(\alpha + \beta j + \gamma_j - \sigma_j)t] \tag{21}$$

The solution of the system of equation (5) is (21) where $\sigma_i = 0$, $i = 1, \ldots,$:

$$b_{ij} = \frac{\alpha + \beta(i-1)}{\gamma_i - \gamma_j + \beta(i-j)} b_{i-1,j};\ i = 1, \ldots, \tag{22}$$

and b_{ii} are determined from the initial conditions in the cells of the channel. The asymptotic solution ($t \to \infty$) is

$$x_i^a = b_{i0} \exp[(\sigma - \alpha - \gamma_0)t] \tag{23}$$

This means that the asymptotic distribution $y_i^a = x_i^a/x^a$ is stationary

$$y_i^a = \frac{b_{i0}}{\sum\limits_{j=0}^{\infty} b_{j0}} \tag{24}$$

regardless of the fact that the amount of substance in the two cells may increase or decrease exponentially. The explicit form of this distribution is

$$y_0^a = \frac{1}{1 + \sum_{i=1}^{\infty} \prod_{k=1}^{i} \frac{\alpha+\beta(k-1)}{\gamma_k-\gamma_0+\beta k}}, \quad y_i^a = \frac{\prod_{k=1}^{i} \frac{\alpha+\beta(k-1)}{\gamma_k-\gamma_0+\beta k}}{\sum_{i=0}^{\infty} \prod_{k=1}^{i} \frac{\alpha+\beta(k-1)}{\gamma_k-\gamma_0+\beta k}}, i = 1, \ldots \tag{25}$$

3 The Model of Two Substances

Let us discuss now a model of moving of two immiscible substances through a channel containing infinite number of cells. The substances enter the channel through the entry cell and in general the following three processes are allowed: the substances may enter the cells one after the another and the substances may be used for some purposes in the corresponding cell. From the point of view of migration flows this model corresponds to migration of migrants with two different values of some characteristics (e.g. different religions).

Let us denote the amount of substance of the two types in the i-th cell of the channel as x_i^1 and x_i^2. The model equations for the movement of the two kings of substance are

$$\begin{aligned} \frac{dx_0^1}{dt} &= \sigma^1 x_0^1 - \alpha^1 x_0^1 - \gamma_0^1 x_0^1; \\ \frac{dx_i^1}{dt} &= [\alpha^1 + \beta^1(i-1)]x_{i-1}^1 - (\alpha^1 + \beta^1 i + \gamma_i^1)x_i^1; \quad i = 1, 2, \ldots \end{aligned} \tag{26}$$

$$\begin{aligned} \frac{dx_0^2}{dt} &= \sigma^2 x_0^2 - \alpha^2 x_0^2 - \gamma_0^2 x_0^2; \\ \frac{dx_i^2}{dt} &= [\alpha^2 + \beta^2(i-1)]x_{i-1}^2 - (\alpha^2 + \beta^2 i + \gamma_i^2)x_i^2; \quad i = 1, 2, \ldots \end{aligned} \tag{27}$$

For the stationary regime of functioning of the channel the amount of the substances and the stationary distributions of the substances in for the two kinds of substances are

$$x_i^{1,*} = \frac{[k^1 + (i-1)]!}{(k^1 - 1)! \prod_{j=1}^{i} (k^1 + j + a_j^1)} x_0^{1,*} \tag{28}$$

$$y_i^{1,*} = \frac{[k^1 + (i-1)]!}{(k^1 - 1)! \prod_{j=1}^{i} (k^1 + j + a_j^1)} y_0^{1,*} \tag{29}$$

$$x_i^{2,*} = \frac{[k^2 + (i-1)]!}{(k^2 - 1)! \prod_{j=1}^{i} (k^2 + j + a_j^2)} x_0^{2,*} \tag{30}$$

$$y_i^{2,*} = \frac{[k^2 + (i-1)]!}{(k^2 - 1)! \prod_{j=1}^{i} (k^2 + j + a_j^2)} y_0^{2,*} \tag{31}$$

where $k^1 = \alpha^1/\beta^1; k^2 = \alpha^2/\beta^2; a_j^1 = \gamma_j^1/\beta^1; a_j^2 = \gamma_j^2/\beta^2; y_i^{1,*} = x_i^{1,*}/x^{1,*}; y_i^{2,*} = x_i^{2,*}/x^{2,*}$ and $x^{1,*}$ and $x^{2,*}$ are the total amounts of the two substances in all cells of the channel.

For the case of non-stationary regime of functioning of the channel the forms of the asymptotic distribution for the two kinds of substances are

$$y_0^{1,a} = \frac{1}{1 + \sum_{i=1}^{\infty} \prod_{k=1}^{i} \frac{\alpha^1 + \beta^1(k-1)}{\gamma_k^1 - \gamma_0^1 + \beta^1 k}}, \quad y_i^{1,a} = \frac{\prod_{k=1}^{i} \frac{\alpha^1 + \beta^1(k-1)}{\gamma_k^1 - \gamma_0^1 + \beta^1 k}}{\sum_{i=0}^{\infty} \prod_{k=1}^{i} \frac{\alpha^1 + \beta^1(k-1)}{\gamma_k^1 - \gamma_0^1 + \beta^1 k}}, i = 1, \ldots \tag{32}$$

$$y_0^{2,a} = \frac{1}{1 + \sum_{i=1}^{\infty} \prod_{k=1}^{i} \frac{\alpha^2 + \beta^2(k-1)}{\gamma_k^2 - \gamma_0^2 + \beta^2 k}}, \quad y_i^{2,a} = \frac{\prod_{k=1}^{i} \frac{\alpha^2 + \beta^2(k-1)}{\gamma_k^2 - \gamma_0^2 + \beta^2 k}}{\sum_{i=0}^{\infty} \prod_{k=1}^{i} \frac{\alpha^2 + \beta^2(k-1)}{\gamma_k^2 - \gamma_0^2 + \beta^2 k}}, i = 1, \ldots \tag{33}$$

4 Discussion

We have mentioned above that the discussed model can be used for study of movement and distribution of migrants in a sequence of countries that form a migration channel. Migration channel of such kind was clearly visible in 2015 when a large influx of migrants in Europe was observed along a channel with Greece as the entry country. The parameters of the models discussed above can be interpreted as follows from the point of view of the application of the models to the migration channels. σ can be considered as a "gate" parameter as it regulates the number of migrants that enter the channel. σ is a parameter that is specific for the entry country of the channel. A small value of σ can decrease significantly the number of migrants that enter the channel. Large value of σ may lead to large migration flows. The value of the parameter σ can be regulated by the authorities of the entry country and by some over-national authorities if such authorities exist and they are allowed to act on the territory of the entry country. Thus if the state structures of the entry country are weak because of some kind of crisis or as a consequences of other reasons then the value of the parameter σ may be large. Thus large number of migrants may enter the channel and this will lead to large problems in the entry country and in all countries along the channel, especially to the countries that are close the entry country and are part of the migration channel.

Another kind of "gate" parameter is the parameter α that regulates the number of migrants that move from one country to the next country in the sequence of countries that form the migration channel. Large value of α means that the movement between the countries is large and the migrants easily cross the state borders. Small value of α means that the the crossing of the borders is more difficult. What was observed for the case of the migration flows in 2015 in Europe was that at some moments the borders have been practically open and for some time the value of the parameter α was large and almost the same for all countries that have been part of the channel. Such kind of situation is modelled by the models presented above.

The parameter β accounts for the attractiveness of the countries of the channel that are distant from the entry country of the channel. The large values of this parameter lead to a tendency for leaving the countries around the beginning of the migration channel and attempts to settle in much more attractive countries along the channel. Parameter γ_i accounts for the number of migrants that enter the i-th country of the channel but do not leave it. The reasons for this may be different, e.g. some migrants may obtain permission to stay in the country. Small values of the parameters γ_i correspond to a large traffic of migrants through the countries of the channel. If in some country the value of the corresponding parameter γ is large then significant number of migrants may stay in this country and the number of the migrant moving further through the channel may decrease.

For the case of two kinds of migrants the number of parameters increase which leads to increasing number of possible situations. For an example the entry country may prefer one of the kinds of migrants and then the values of the "gate" parameters g^1 and g^2 may have quite different values. The countries along the channel may impose different level of difficulty of crossing the borders which may lead in differences in the values of the parameters α^1 and α^2. The countries along the channel may have different level of attractiveness which will lead to different values of the parameters β^1 and β^2. Finally different countries may have different preferences about the numbers and about the kind of migrants they allow to stay in the country. This will lead to different values of the parameters γ_i^1 and γ_i^2.

Let us discuss the above remarks in more detail by means of the obtained mathematical results. Let us first consider the case of one kind of migrants and stationary regime of functioning of the channel. From Eq. (10) we obtain

$$\frac{y_i^*}{y_{i+1}^*} = 1 + \frac{\beta + \gamma_{i+1}}{\alpha + \beta i} \tag{34}$$

Equation (34) shows us that for the case of infinite channel the number of migrants has to decrease with increasing value of i, i.e., in the countries that are away from the entry country of the channel the number of migrants in any country is smaller than the number of migrants in the previous country of the channel. Then for the case of infinite channel the effect of concentration of migrants in countries that are far from the entry country is not observed even if these countries are very attractive to the migrants. Let us note here that the models considered above can be extended to the case of a channel containing finite number of cells (finite number of countries for the vase of migrant flows). Then a new effect will be observed: concentration of migrants in the last cell of the channel (the final destination country). If the attractiveness of the final destination country is large then the concentration of the migrants will be observed in the entry country of the channel and in the final destination country. Such possible developments of the situation in the migration channels will be discussed elsewhere.

Equation (34) shows additional details about the influence of the parameters of the channel on the migrant flow. The increase of α leads to decreasing of the ratio y_i^*/y_{i+1}^*. This means that the larger value of the gate parameter α leads to a smoothing of the distribution of the number of migrants along the countries of the channel. Thus if the countries that are far from the entry country of the channel want to decrease the number of migrants on its own territory the have to take measures to decrease the value of the parameter α. The increasing of the value of the parameter β has more complicated influence on the ratio y_i^*/y_{i+1}^* but in principle the effect is the same as the increasing of the value of parameter α. If the countries that are attractive for migrants have a policy to decrease their attractiveness then smaller number of migrants will reach their territory. If such countries have "open doors" politics towards migrants then larger number of migrants will leave the countries around the entry country of the channel and will move towards the attractive countries. In combination with large value of σ this may lead to floods of migrants in the attractive countries. But this will lead also to even larger flood of migrants in the countries that are around the entry country of the channel. This was observed indeed for the case of the massive migration in Europe in 2015. Finally the increasing value of the "leakage" parameter γ_{i+1} will lead to decreasing of the ratio y_i^*/y_{i+1}^*. The reason for this is obvious: when more migrants change their migrant status (e.g. obtain permission to stay) in the the $i+1$-th country then the number of migrants without status will decrease and this will lead to decreasing ratio y_i^*/y_{i+1}^*.

Let us now discuss further the case of two kinds of migrants for the stationary regime of the functioning of the channel. In this case the ratio of the numbers of migrants in a cell of the channel is

$$\frac{x_i^{1,*}}{x_i^{2,*}} = \frac{x_0^{1,*}}{x_0^{2,*}} \frac{[k^1 + (i-1)]!}{[k^2 + (i-1)]!} \frac{(k^2-1)!}{(k^1-1)!} \frac{\prod_{j=1}^{i}(k^2 + j + a_j^2)}{\prod_{j=1}^{i}(k^1 + j + a_j^1)} \quad (35)$$

Equation (35) leads to the following conclusions. First of all the mix of the migrants in any country of the channel depends on the ratio $x_0^{1,*}/x_0^{2,*}$ at the entry country of the channel. Thus the politics in the entry country of the channel towards the different categories of migrants is extremely important for the distribution of migrants in the entire channel. If there are migrants with unfavorable characteristics from the point of view of countries from the the channel then one of most effective ways to reduce the number of such migrants is to reduce their possibility to enter the channel. The influence of the other parameters of the channel is more complicated. Large values of γ_i^2 with respect to the values of γ_i^1 (this corresponds, e.g., for much larger acceptance of the migrants of class 2 in comparison to the migrants of class 1 in all countries of the channel up to the i-th country of the channel) when the other parameters of the channel are equal for the two classes of migrants lead to larger value of the ratio $x_0^{1,*}/x_0^{2,*}$. This means that the mix of migrants without status in the i-th country of the channel depends on the politics of previous countries of the channel towards different categories of migrants. Finally the ratio of the migrants depends on the parameters k^1 and k^2 that are ratios α^1/β^1 and α^2/β^2, i.e. the ratios between corresponding possibility for mobility of the class of migrants (the parameters α^1 and α^2) and the attractiveness of the countries far away from the entry country of the channel (the parameters β^1 and β^2). This dependence is the most complicated one. If the ratios k^1 and k^2 are the same then the influence of k^1 and k^2 adds nothing to the influence of parameters a^1 and a^2 on the mix of migrants in the countries of the channel. In order to obtain more information about the influence of $k^{1,2}$ (respectively about the influence of the parameters $\alpha^{1,2}$ and $\beta^{1,2}$ we can use the recurrence relationships for the numbers of migrants in the i-th country of the channel. These relationships may be written as

$$x_i^{1,*} = \frac{1}{1 + \frac{\beta^1 + \gamma_i^1}{\alpha^1 + \beta^1(i-1)}} x_{i-1}^{1,*}, \; x_i^{2,*} = \frac{1}{1 + \frac{\beta^2 + \gamma_i^2}{\alpha^2 + \beta^2(i-1)}} x_{i-1}^{2,*} \quad (36)$$

Then the vary large values of the parameter β increase the traffic of migrants through the channel and the decrease of the number of migrants in the i in comparison to the number of migrants in $i-1$-th country will be approximately $1 - 1/i$ as a proportion. The influence of the very large values of the parameters α is larger. When the value of the parameter α is much large that the values of the corresponding parameters β and γ_i then the number of migrants in the i-th country of the channel will be approximately equal to the number of migrants in the i-th country of the channel. Thus the decreasing permeability of the borders between the neighbouring states from the migration channel is more effective that decreasing of the attractiveness of the countries far from the entry country of the channel.

References

1. Armitage, R.: Population projections for English local authority areas. Popul. Trends **43**, 31–40 (1986)
2. Borjas, G.J.: Economic theory and international migration. Int. Migr. Rev. **23**, 457–485 (1989)
3. Bracken, I., Bates, J.J.: Analysis of gross migration profiles in England and Wales: some developments in classification. Environ. Plann. A **15**, 343–355 (1983)
4. Champion, A.G., Bramley, G., Fotheringham, A.S., Macgill, J., Rees, P.H.: A migration modelling system to support government decision-making. In: Stillwell, J., Geertman, S. (eds.) Planning Support Systems in Practice, pp. 257–278. Springer, Berlin (2002)
5. Chen, W.-C.: On the weak form of the Zipf's law. J. Appl. Probab. **17**, 611–622 (1980)
6. Diodato, V.: Dictionary of Bibliometrics. Haworth Press, Binghampton, NY (1994)
7. Ethier, W.J.: International trade and labor migration. Am. Econ. Rev. **75**, 691–707 (1985)
8. Fawcet, J.T.: Networks, linkages, and migration systems. Int. Migr. Rev. **23**, 671–680 (1989)
9. Greenwood, M.J.: Modeling migration. In: Kemp-Leonard, K. (ed.) Encyclopedia of Social Measurement, vol. 2, pp. 725–734. Elsevier, Amsterdam (2005)
10. Gurak, D.T., Caces, F.: Migration networks and the shaping of migration systems. In: Kitz, M.M., Lim, L.L., Zlotnik, H. (eds.) International Migration Systems: A Global Approach, pp. 150–176. Clarendon Press, Oxford (1992)
11. Harris, J.R., Todaro, M.P.: Migration, unemployment and development: a two-sector analysis. Am. Econ. Rev. **60**, 126–142 (1970)
12. Hotelling, H.: A mathematical theory of migration. Environ. Plan. **10**, 1223–1239 (1978)
13. Irwin, J.O.: The place of mathematics in medical and biological sciences. J. Roy. Stat. Soc. **126**, 1–44 (1963)
14. Irwin, J.O.: The generalized waring distribution applied to accident theory. J. Roy. Stat. Soc. **131**, 205–225 (1968)
15. Ledent, J.: Multistate life table: movement versus transition perspectives. Environ. Plan. A **12**, 533–562 (1980)
16. Lee, E.S.: A theory of migration. Demography **3**, 47–57 (1966)
17. Massey, D.S., Arango, J., Hugo, G., Kouaougi, A., Pellegrino, A., Edward, Taylor J.: Theories of international migration: a review and appraisal. Popul. Dev. Rev. **19**, 431–466 (1993)
18. Puu, T.: A simplified model of spatiotemporal population dynamics. Environ. Plan. **17**, 1263–1269 (1985)
19. Puu, T.: Hotelling's migration model revisited. Environ. Plan. **23**, 1209–1216 (1991)
20. Schubert, A., Glänzel, W.: A dynamic look at a class of skew distributions. A model with scientometric application. Scientometrics **6**, 149–167 (1984)
21. Simon, H.A.: On a class of skew distribution functions. Biometrica **42**, 425–440 (1955)
22. Simon, J.H.: The Economic Consequences of Migration. The University of Michnigan Press, Ann Arbor, MI (1999)
23. Skeldon, R.: Migration and Development: A Global Perspective. Routledge, London (1992)
24. Vitanov, N.K.: Science Dynamics and Research Production: Indicators, Indexes, Statistical Laws and Mathematical Models. Springer International (2016)
25. Vitanov, N.K., Vitanov, K.N.: Population dynamics in presence of state dependent fluctuations. Comput. Math. Appl. **68**, 962–971 (2013)
26. Vitanov, N.K., Vitanov, K.N.: Box model of migration channels. Math. Soc. Sci. **80**, 108–114 (2016)
27. Vitanov, N.K., Ausloos, M., Rotundo, G.: Discrete model of ideological struggle accounting for migration. Adv. Complex Syst. **15**(1) (2012). Article number 1250049
28. Vitanov, N.K., Dimitrova, Z.I., Ausloos, M.: Verhulst-Lotka-Volterra model of ideological struggle. Physica A **389**, 4970–4980 (2010)
29. Vitanov, N.K., Dimitrova, Z.I., Vitanov, K.N.: Traveling waves and statistical distributions connected to systems of interacting populations. Comput. Math. Appl. **66**, 1666–1684 (2013)

30. Vitanov, N.K., Jordanov, I.P., Dimitrova, Z.I.: On nonlinear dynamics of interacting populations: coupled kink waves in a system of two populations. Commun. Nonlinear Sci. Numer. Simul. **14**, 2379–2388 (2009)
31. Vitanov, N.K., Jordanov, I.P., Dimitrova, Z.I.: On nonlinear population waves. Appl. Math. Comput. **215**, 2950–2964 (2009)
32. Weidlich, W., Haag, G. (eds.): Interregional Migration: Dynamic Theory and Comparative Analysis. Springer, Berlin (1988)
33. Willekens, F.J.: Probability models of migration: complete and incomplete data. SA J. Demogr. **7**, 31–43 (1999)
34. Willekens, F.: Models of migration observations and judgement. In: Raymer, J., Willekens, F. (eds.) International Migration in Europe: Data, Models and Estimates, pp. 117–147. Wiley, New York (2008)

Modified Coordinates in Dynamics Simulation of Multibody Systems with Elastic Bodies

Evtim V. Zahariev

Abstract In the paper, modified coordinates are applied for simulation of large flexible deflections of elastic multibody systems. Absolute finite element and joint coordinates are proposed. The joint coordinate positions are defined solving non-linear programming problem for which the general function is the quadratic form of the minimal node deviations of the beam model with respect to the real large deviations. So, with minimal number of coordinates the nonlinear effects in case of large elastic deflections could be approximated. The novelty in the paper is an approach to optimization and effective application of the broadly applied absolute and joint coordinates in dynamics simulation of flexible structures. These coordinates are called modified since differs in location and computation of the kinematic parameters and corresponding stiff and mass properties of the links. Both coordinates significantly simplifies the computer code generation of the dynamic equations and diminish the computations. Examples of flexible deflection of large flexible beam and a four stroke flexible structure are presented.

1 Introduction

Theoretical basis of the Finite Element Theory (FET) for definition of deflection coordinates and deriving the mass and material properties of the discretized elastic bodies has been applied in many commercial software packages. It was proven [4] that using the FET approach the dynamic equations are free of centrifugal and Coriolis accelerations and the corresponding inertia forces. Using the classical FET approach the so called stiffening effects in non-isoparametric elements is difficult to be taken into account.

To illustrate the problem the simple non-isoparametric beam element with two degree of freedom of each node is discussed. In Fig. 1 the beam element is presented

E.V. Zahariev (✉)
Institute of Mechanics, Bulgarian Academy of Sciences,
Acad. G. Bonchev St., bl. 4, 1113 Sofia, Bulgaria
e-mail: evtimvz@bas.bg

K. Georgiev et al. (eds.), *Advanced Computing in Industrial Mathematics*, Studies in Computational Intelligence 728,
https://doi.org/10.1007/978-3-319-65530-7_19

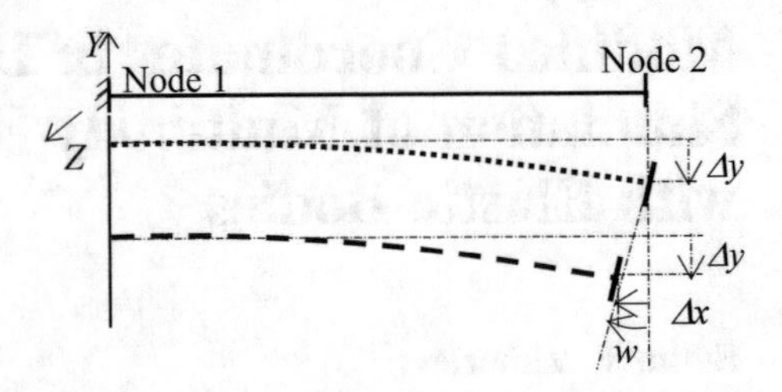

Fig. 1 Flexible node deflections in FET and the real deflections

with its possible deflections along the coordinate system axes Y and Z. In the figure, with dotted line, the idealized FET deflections are depicted where only transversal Δy and the angular w deflections are used. Actually, this assumption is valid for small deflections only. With dashed line the actual deflections are shown where, because of the transversal and angular deflections, node 2 experience additional deflection $\Delta x = \Delta x\,(\Delta y, w)$. That is the so-called stiffening effect that is not taken into account in case of classical FET and some of the commercially available software.

Regarding the velocity dependent terms in the dynamic equations (centrifugal forces and Coriolis' accelerations) the inertia forces defined by the FET are as follows [1]:

$$\mathbf{F} = \mathbf{M} \cdot \ddot{\mathbf{\Delta}} \tag{1}$$

where $\mathbf{F}$ is $n \times 1$ matrix vector of the inertia forces, $\mathbf{M}$ is $n \times n$ global mass matrix and $\ddot{\mathbf{\Delta}}$ is $n \times 1$ matrix vector of the flexible node deflections. Although the flexible elements are more complex elastic bodies than the rigid bodies, no velocity dependent terms are taken into account.

Recently a novel method of ANCF has been developed [8]. It is based on the theory of the curvilinear coordinates. The slopes of the nodes in the tangential line or plane of the space curve or shell are applied as coordinates. The method proposes important advantages mainly in large rotation simulation and mass matrices formulation. Review on the ANCF method for large deformation dynamics simulation of flexible multibody systems was presented by Gerstmayr et al. [3]. The paper provided a comprehensive review of the advantages and further developments of the ANCF method.

Khan and Anderson [5] proposed the most recent classification of the methods for MBS dynamics. They applied the Floating Frame of Reference (FFR) and Absolute Nodal Coordinate Formulation (ANCF) approaches for realization of their numerical procedure. The FFR [4] approach was developed and successfully applied for dynamics simulation of spatial motion and rotation of flexible multibody systems. This approach is also used for application of the modal coordinates and significant decreasing of the flexible coordinates.

Meijard [6] derived generalized Newton—Euler dynamic equations for the case of FFR formulation. The equations represent the inertia forces as function of the velocities and accelerations of the global and flexible coordinates. Zahariev [9] derived generalized Newton-Euler dynamics equations for the rigid and flexible bodies for which the kinetic energy is quadratic form of the velocities. The equations are applicable for the flexible elements of the FET. The inertia forces are with respect to the quasi velocities and accelerations and are invariant to the kind of the coordinates.

In [10] a method of Finite Elements in Relative Coordinates (FERC) based on the FET was proposed for dynamic simulation of large flexible structures. Using the generalized Newton-Euler dynamic equations nonlinear effects and velocity dependent terms are successfully simulated. But using this approach for complex structures with many mutually connected adjacent flexible bodies one could experience a lot of difficulties that mainly consists in its program system realization and complex pre-processor procedures.

A method of Finite Elements in Absolute Coordinates (FEAC) is discussed in the paper. It was presented in [10] and further developed for different applications. FEAC with application of modified absolute nodal coordinates is proposed here for simulation of structures subject to seismic excitations.

For taking into account the stiffening effects in elastic beam elements modified joint coordinates are discussed and the possible flexible deflections of the beam are presented by two rotational joint coordinates. The problem for selection of the modified joint coordinates and their position is defined as an optimization problem minimizing the node deviations of the beam model with respect to the real large scale deviations.

2 Large Spatial Rotations of Flexible Elements and Modified Absolute Nodal Coordinates

To understand the difference between the classical FET and the method of the finite elements in absolute nodal coordinates using multibody system methodology some explanations about the kinematics of flexible nodes will be presented here.

In Fig. 2 a non-isoparametric space beam element (with index i) is presented. This finite element is relatively simple for explanation but, at the same time, presents the six degree of freedom (dof) space deflections of the nodes. In the figure the beam and small translations and rotations, respectively $\mathbf{\Delta}_{i,k} = [\,\mathbf{\Delta}x_{i,k} \quad \mathbf{\Delta}y_{i,k} \quad \mathbf{\Delta}z_{i,k}\,]^{\backslash}$, $\boldsymbol{\theta}_{i,k} = [\,\theta x_{i,k} \quad \theta y_{i,k} \quad \theta z_{i,k}\,]^{\backslash}$, $k = 1, 2$ of its nodes (indices i, 1 and i, 2—the first and second nodes of element i) with respect to the initial not deformable element coordinate system $\underline{X}_i\underline{Y}_i\underline{Z}_i$ are depicted. With the superscript "\" a matrix transpose is denoted. The underlined notations denote initial configurations. The matrices are pointed out by bold characters. For scalars and objects (for example points, bodies, coordinate systems) italic notations are used.

It should be said that in FET these deflections are with respect to the element reference frame in its initial position (not deformed position) and no rotation of the element coordinate system is taken into account. Of course, coordinate system transformation with respect to the absolute reference frame $X_0Y_0Z_0$ of mass and

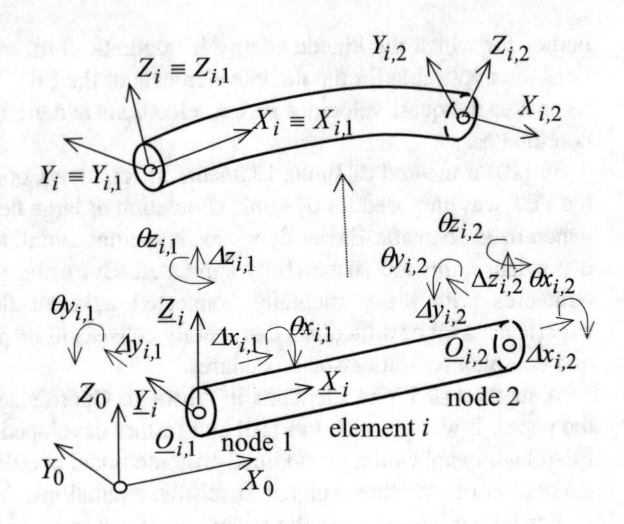

Fig. 2 Small flexible node deflections of space non-isoparametric beam finite element

stiffness matrices is considered but for the structure initial configuration only. As a result of the node deflections the element coordinate system $X_iY_iZ_i$, respectively the node i, 1 coordinate system $X_{i,1}Y_{i,1}Z_{i,1}$, moves to a new position defined by the radius—vector $\boldsymbol{\rho}_{i,1}$ that could be easily estimated adding the initial position of element coordinate system origin, radius—vector $\underline{\boldsymbol{\rho}}_i \equiv \underline{\boldsymbol{\rho}}_{i,1}$, and the node i, 1 deflections $\boldsymbol{\Delta}_{i,1}$. The same holds for the radius–vector $\boldsymbol{\rho}_{i,2}$ taking into account the initial position of node i, 2 coordinate system origin, radius–vector $\underline{\boldsymbol{\rho}}_{i,2}$, and the node i, 2 deflections $\boldsymbol{\Delta}_{i,2}$. The node deflections of the *i-th* beam element are set in a 12×1 matrix—column $\boldsymbol{\Theta}_i = \left[\boldsymbol{\Theta}_{i,1}^{\backslash} \quad \boldsymbol{\Theta}_{i,2}^{\backslash}\right]^{\backslash} = \left[\boldsymbol{\Delta}_{i,1}^{\backslash} \quad \boldsymbol{\theta}_{i,1}^{\backslash} \quad \boldsymbol{\Delta}_{i,2}^{\backslash} \quad \boldsymbol{\theta}_{i,2}^{\backslash}\right]^{\backslash}$.

In Fig. 3 the beam element in the configuration as this of Fig. 2 is depicted including the length L_i of the beam (along axis X) and the small deflections $\boldsymbol{\Delta}_{i,2\rangle i}$ of node i, 2 relative to the beam coordinate system i. In the right subscript $\boldsymbol{\Delta}_{i,2\rangle i}$ the symbol "⟩" serves as an arrow pointing out that the coordinate system with index i, 2 is with respect to coordinate system i.

The right symbol of the subscript could be omitted if it points the absolute reference frame with index "0"—zero. Obviously, the beam configuration is quite the same and the fact that the flexible deflections are read relative to the moving

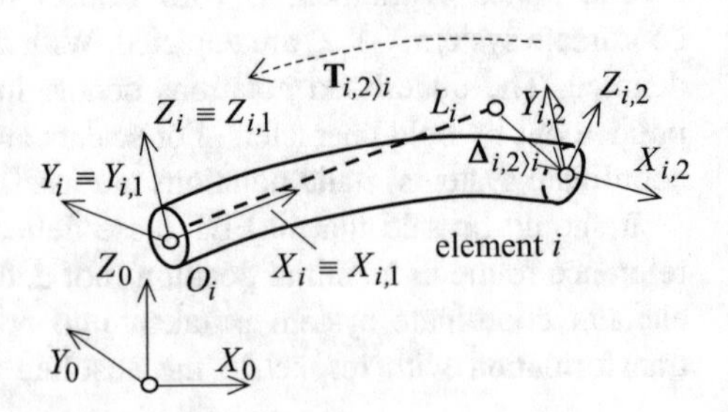

Fig. 3 Finite beam element in relative coordinates

element coordinate system does not change the regulations for FET discretization and for computation of the mass and stiffness matrices. The position of node i, 2 coordinate system relative to the coordinate system of the moving beam is presented by 4×4 homogeneous transformation matrix $\mathbf{T}_{i,2\rangle i}$, i.e.:

$$\mathbf{T}_{i,2\rangle i} = \begin{bmatrix} {}^{3.3}\boldsymbol{\tau}_{i,2\rangle i} & {}^{3,1}\boldsymbol{\rho}_{i,2\rangle i} \\ {}^{1,3}\mathbf{0} & 1 \end{bmatrix} = \begin{bmatrix} 1 & -\theta z_{i,2\rangle i} & \theta y_{i,2\rangle i} & L_i + \Delta x_{i,2\rangle i} \\ \theta z_{i,2\rangle i} & 1 & -\theta x_{i,2\rangle i} & \Delta y_{i,2\rangle i} \\ -\theta y_{i,2\rangle i} & \theta x_{i,2\rangle i} & 1 & \Delta z_{i,2\rangle i} \\ 0 & 0 & 0 & 1 \end{bmatrix} = \begin{bmatrix} 1 & -q_{i_6} & q_{i_5} & L + q_{i_1} \\ q_{i_6} & 1 & -q_{i_4} & q_{i_2} \\ -q_{i_5} & q_{i_4} & 1 & q_{i_3} \\ 0 & 0 & 0 & 1 \end{bmatrix} \tag{2}$$

which for small rotational deflections does not include trigonometric functions. With this assumption that the homogeneous transformation matrix is linear for the relative coordinates q_{i_j}, $j = 1, 2, \ldots, 6$ of element i that include the small translational and rotational flexible deflections. ${}^{3,3}\boldsymbol{\tau}_{i,2\rangle i}$ is 3×3 rotational matrix; the left superscripts point out (used if needed) the matrix dimensions (rows, columns).

In Fig. 4, similar to Fig. 2, the coordinate systems of nodes i, 1 and i, 2, $X_{i,k}Y_{i,k}Z_{i,k}$; $i = 1, 2$, are located in their final configuration. The initial positions of the node coordinate systems are presented by the same but underlined notations. FET claims for the small deflections to be read with respect to the system initial configuration.

The homogeneous transformation matrices $\mathbf{T}_{i,1\rangle 0}$, $\mathbf{T}_{i,2\rangle 0}$ of the absolute position of the node coordinate systems i, 1 and i, 2, respectively, are expressed by the matrices $\underline{\mathbf{T}}_{i,1\rangle 0}$ and $\underline{\mathbf{T}}_{i,2\rangle 0}$ of their initial configuration and the transformation matrix because of the small node deflections.

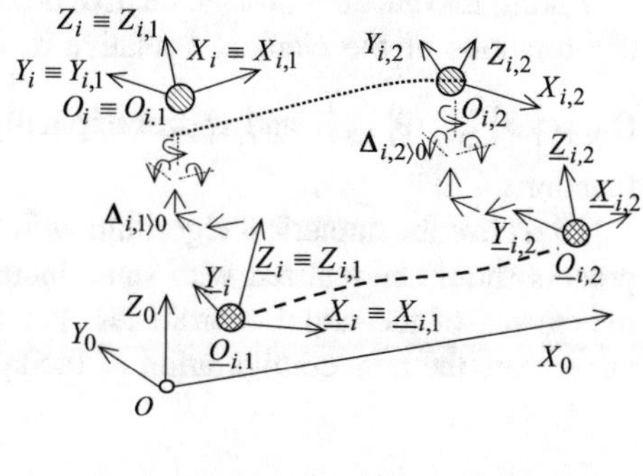

Fig. 4 Nodes as free coordinate systems and corresponding motion coordinates

$$\mathbf{T}_{i,k\rangle 0} = \begin{bmatrix} \boldsymbol{\tau\Delta}_{i,k\rangle 0} \cdot \boldsymbol{\tau}_{i,k} & \underline{\boldsymbol{\rho}}_{i,k} + \boldsymbol{\Delta}_{i,k} \\ {}^{1,3}\mathbf{0} & 1 \end{bmatrix} = \begin{bmatrix} & \tau\Delta_{i,k\rangle 0} \cdot \underline{\boldsymbol{\tau}_{i,k}} & & \underline{\rho x}_{i,k} + q_{i,k_1} \\ & & & \underline{\rho y}_{i,k} + q_{i,k_2} \\ & & & \underline{\rho z}_{i,k} + q_{i,k_3} \\ 0 & 0 & 0 & 1 \end{bmatrix};$$

$$\boldsymbol{\Delta}^{\backslash}_{i,k\rangle 0} = [\, q_{i,k_1} \quad q_{i,k_2} \quad q_{i,k_3} \,]^{\backslash};$$

$$\boldsymbol{\tau\Delta}_{i,k\rangle 0} = \begin{bmatrix} 1 & -\theta z_{i,k\rangle 0} & \theta y_{i,k\rangle 0} \\ \theta z_{i,k\rangle 0} & 1 & -\theta x_{i,k\rangle 0} \\ -\theta y_{i,k\rangle 0} & \theta x_{i,k\rangle 0} & 1 \end{bmatrix}, \; k = 1, 2, \tag{3}$$

where $\mathbf{q}_{i,k} = \begin{bmatrix} \boldsymbol{\Delta}^{\backslash}_{i,k\rangle 0} & \boldsymbol{\theta}^{\backslash}_{i,k\rangle 0} \end{bmatrix}^{\backslash} = [q_{i,k_1} \quad q_{i,k_2}, \ldots, q_{i,k_6}]^{\backslash}$, $k = 1, 2$ are the deflections of node k from element i with respect to the absolute reference frame (index 0), which are also the coordinates of the free moving nodes. To apply Eq. (2) the deflections are to be small although they do not present the flexible deformations only. The deflections q_{i,k_n}, $k = 1, 2$; $n = 1, 6$ include also the global motion of the nodes of element i. To define the relative flexible deformations of the element i the relative position of node i, 2 with respect to node i, 1 that coincides with the element coordinate system i, should be defined, i.e.:

$$\begin{aligned} \mathbf{T}_{i,2\rangle i,1} &= \mathbf{T}^{-1}_{i,1\rangle 0} \cdot \mathbf{T}_{i,2\rangle 0} = \left[\underline{\mathbf{T}}_{i,1\rangle 0} \cdot \mathbf{T\Delta}_{i,1}\right]^{-1} \cdot \underline{\mathbf{T}}_{i,2\rangle 0} \cdot \mathbf{T\Delta}_{i,2} \\ &= \mathbf{T\Delta}^{-1}_{i,1} \underline{\mathbf{T}}^{-1}_{i,1\rangle 0} \cdot \underline{\mathbf{T}}_{i,2\rangle 0} \cdot \mathbf{T\Delta}_{i,2} \end{aligned} \tag{4}$$

The matrix elements $\mathbf{T}_{i,2\rangle i,1} = \mathbf{T}_{i,2\rangle i}$ are actually the flexible deformations (rotations and translations) and, since they are to be small, the rotations could be presented without trigonometric functions, i.e.:

$$\mathbf{T}_{i,2\rangle i} = \begin{bmatrix} 1 & -\theta z_{i,2\rangle i} & \theta y_{i,2\rangle i} & L_i + \Delta x_{i,2\rangle i} \\ \theta z_{i,2\rangle i} & 1 & -\theta x_{i,2\rangle i} & \Delta y_{i,2\rangle i} \\ -\theta y_{i,2\rangle i} & \theta x_{i,2\rangle i} & 1 & \Delta z_{i,2\rangle i} \\ 0 & 0 & 0 & 1 \end{bmatrix} \tag{5}$$

Taking into account that the relative deformations of node 1 are zero, the flexible deformations of the element i relative to its own reference frame are as follows $\boldsymbol{\Theta}_{i\rangle i} = \begin{bmatrix} \boldsymbol{\Delta}^{\backslash}_{i,2\rangle i} & \boldsymbol{\theta}^{\backslash}_{i,2\rangle i} \end{bmatrix}^{\backslash}$ and appear explicitly in matrix $\mathbf{T}_{i,2\rangle i}$ without trigonometric functions.

To realize the numerical algorithm with the proposed coordinates the integration process should be realized with small increments of time, respectively, with small increments of the nodal coordinates. For this purpose on every step of the time increments the new configuration of the system is the initial configuration for the next iteration and the absolute nodal coordinates start again with zero values. That process is not disadvantage since during the numerical integration of the dynamic equations the new positions of the coordinate systems are to be recalculated and the new mass and stiffness properties of the bodies are to be computed. Detailed statements of the procedures are presented in [11].

3 Modified Joint Coordinates of Beam Flexible Deflections

In Fig. 5 the beam element that experiences small deflections in *XY* plane is divided into three smaller insubstantial beam segments connected by two rotational joints. In the joints linear springs are set up that cause torque in case of motion (bending). The sum of the segment lengths is equal to the beam element length. The transformation between rotations q_1, q_2 in the joints and deflections in node 2 is derived assuming small rotations q_1, q_2, i.e.:

$$q_1 + q_2 = \theta z_2; \ (l_1 + l_2).q_1 + l_2.q_2 = \Delta y_2 \tag{6}$$

which results to a new matrix of the elasticity that should be equal to the stiffness matrix $\mathbf{L}_{xy}^2$ in FET, i.e.:

$$\mathbf{L}_{xy}^2 = \begin{bmatrix} \frac{L^3}{3.E.I_z} & \frac{L^2}{2.E.I_z} \\ \frac{L^2}{2.E.I_z} & \frac{L}{E.I_z} \end{bmatrix} = \begin{bmatrix} c_1 . (l_1 + l_2)^2 + c_2 . l_2^2 & c_1 . (l_1 + l_2) + c_2 . l_2 \\ c_1 . (l_1 + l_2) + c_2 . l_2 & c_1 + c_2 \end{bmatrix} \tag{7}$$

Equation 7 is transformed to equations system of three nonlinear equations with respect to four un known—l_1, l_2, q_1, q_2, i.e.:

$$l_1 = \frac{L^2/3 + l_2^2 - L.l_2}{L/2 - l_2} \tag{8}$$

$$c_1 = \frac{1}{E.I_z} \cdot \frac{L^2/2 - L.l_2}{l_1} \tag{9}$$

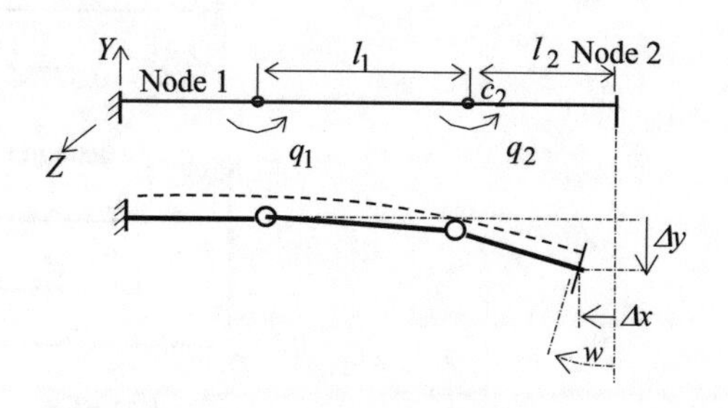

Fig. 5 Kinematic model of bending beam element

$$c_2 = \frac{L}{E.I_z} - c_1 \tag{10}$$

The boundary solutions, for which the joints are set up between the two nodes, are in the interval $0 \le l_2 \le 1/3.L$. For these values of l_2 the values of l_1 are in the interval $0.57735.L \le l_1 \le 2/3.L$.

The boundary solutions, for which the joints are set up between the two nodes, are in the interval $0 \le l_2 \le 1/3.L$. For these values of l_2 the values of l_1 are in the interval $0.57735.L \le l_1 \le 2/3.L$. It is seen that if $l_2 = \frac{L-l_1}{2} = \frac{L-(0.57735).L}{2} = (0.211325).L$ so $c_1 = c_2 = \frac{L}{2.E.I_z}$ and there is no possible solution $c_1 = c_2 = \frac{L}{2.E.I_z}$. It could be seen that there is no possible solution $l_1 = 2.l_2$ or $l_1 = l_2$ as it is discussed in [11]. In Fig. 6 the cases when $c_1 = c_2$ and another practical solution with fractions (Fig. 6) are shown.

An example of large flexible beam in its different positions of deformation is shown in Fig. 7. The size and stiffness properties are depicted in the figure. This example is a comparison to the results obtained in [7] and is presented here to proof the approach proposed here. Using 10 element discretization and the modified joint coordinates, excellent agreement with the experimental results of the deformations are obtained (the dotted lines). For loading torque of 2226.504 Nm the beam implements deformation as a pure circle. For comparison the same example is solved using the classical FET. With dense line the results of deformations are presented for which significant deviations from the experiments are obtained. It could be seen that if the same torque is applied the results does not correspond to the experiments.

In Fig. 8 another example is presented that demonstrate the possibility for successive application of the modified joint coordinates. One element flexible beam is presented which deflections are computed using only two modified joint coordinates and the extremely large flexible deflections (dashed lines) are compared to

Fig. 6 Recommended kinematic models of elastic beam elements

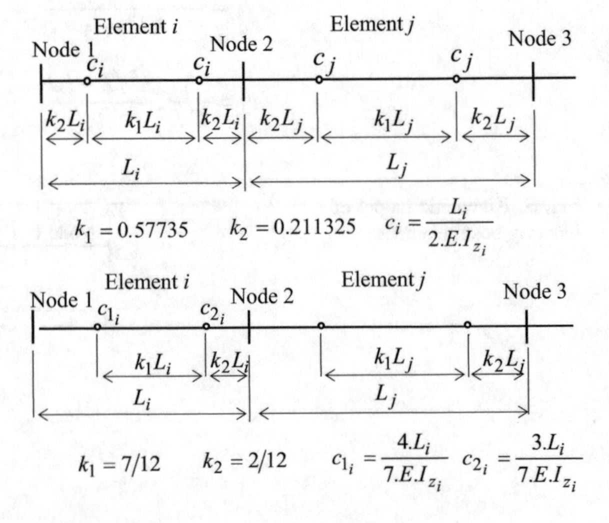

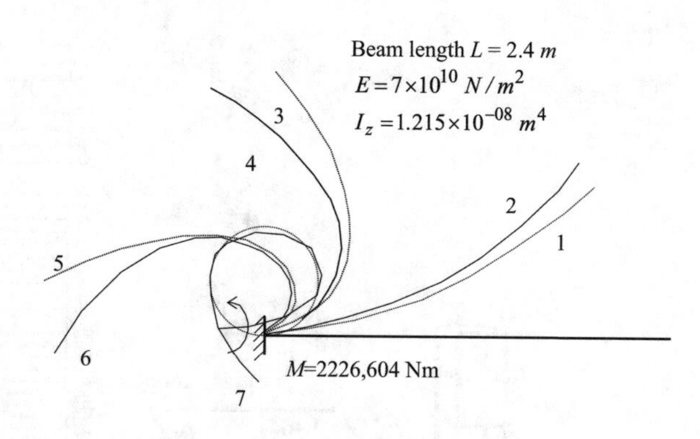

Fig. 7 Large flexible deflections of slender beam

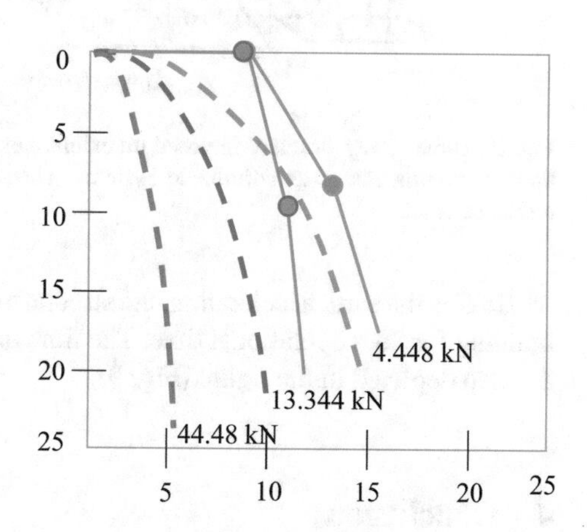

Fig. 8 Large flexible deflections of single flexible element. Beam length $L = 2.4$ m; $EI = 516.54 \times 103$ N m^2

the results obtained in [2] using nonlinear theory. The beam is loaded by vertical forces which values are depicted in the figure. It could be seen that even for large deformations the method prosed in the paper could be applied obtaining admissible results. This could significantly decrease the number of the system coordinates which leads to successive solutions. For extremely high forces (44.48 kN, Fig. 8) the method could not be applied.

In Fig. 9 an example of space flexible structure compiled of rigidly connected beams that build many closed contours is presented. The structure is mounted on a basement imposed on external excitation—one dimensional translational motion q_1 in the horizontal plane and one dimensional rotational excitation q_2 along the vertical axis. Because of the external excitations the basement implements two dimensional motion which functions of the accelerations are $\ddot{q}_1 = a_1 \cdot b_1 \cdot \cos(b_1 \cdot t)$; $\ddot{q}_2 = a_2 \cdot b_2 \cdot \cos(b_2 \cdot t)$. The nodes of the beams are shown as moving coordinate systems which origins are pointed out as O_i, $i = 1, 2, \ldots, 16$, while their axes are parallel to the absolute reference frame. Modified absolute nodal coordinates (Sect. 2) are applied. The nodes 1–4 are connected to the basement and are considered part of a rigid body, i.e.: their motion characteristics coincide with that of the basement. For the following values of the coefficients $a_1 = 0.2 \cdot G$; $b_1 = \pi$; $a_2 = 0.04 \cdot G$; $b_2 = 1.5 \cdot \pi$,

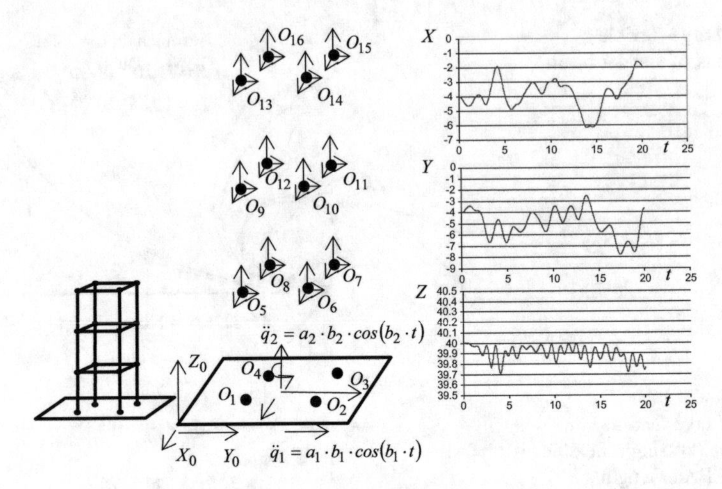

Fig. 9 Three storey structure imposed on external excitation (earthquake) and its dynamic model as free moving nodes as coordinate systems. Deviations of node 16 with respect to absolute reference frame

where G is the earth acceleration, the structure deflections do not cause collapse of the building for 20 s operational time. The time histories of the deformations of node 16 are also depicted in the figure (Fig. 9).

4 Conclusion

Methods of modified absolute nodal coordinates and of modified joint coordinates for flexible elements is proposed and successfully applied for static and dynamic simulation of beam structures. The methods enable significant decreasing of coordinates and simplification of static and dynamic equations.

Acknowledgements The author acknowledges the financial support of Ministry of Education and Science, contract No. DUNC 01/3—2009.

References

1. Doyle, J.F.: Nonlinear Analysis of Thin—Walled Structures. Springer (2001)
2. Fertis, D.G.: Nonlinear Structural Engineering with Unique Theories and Methods to Solve Effectively Complex Nonlinear Problems. Springer (2006)
3. Gerstmayr, J., Sugiyama, H., Mikkola, A.: Review on the absolute nodal coordinate formulation for large deformation analysis of multibody systems. ASME J. Comput. Nonlinear Dyn. **8**(3) (2013)

4. Kane, T.R., Ryan, R.R., Banerjee, A.R.: Dynamics of a cantilever beam attached to a moving rigid base. AIAA J. Guidance Control Dyn. **10**, 139–151 (1987)
5. Khan, I. M., Anderson, K. S.: Large deformation formulation for the flexible divide-and-conquer algorithm. In: Proceedings of the ECCOMAS Thematic Conference Multibody Dynamics, pp. 343–357, University of Zagreb. Croatia (2013)
6. Meijard, J.P.: Validation of flexible beam element in dynamics programs. Nonlinear Dyn. **9**(1–2), 21–36 (1996)
7. Schwertassek, R.: Flexible bodies in multibody systems. In: Angeles, J., Zakhariev, E., (eds.) Computational Methods in Mechanical Systems: Mechanism Analysis, Synthesis and Optimization, NATO ASI, Series F, vol. 161, pp 329–363, Springer, Berlin (1998)
8. Shabana, A.A.: An absolute nodal coordinate formulation for the large rotation and deformation analysis of flexible bodies. Technical Report MBS96-1-UIC, Department of Mechanical Engineering. University of Illinois at Chicago (1996)
9. Zahariev, E.V.: Relative Finite Element Coordinates in Multibody System Simulation. Multibody Syst. Dyn. **7**(1), 51–77 (2002)
10. Zahariev, E.V.: Generalized finite element approach to dynamics modeling of rigid and flexible systems. Mech. Based Des. Struct. Mach. **34**(1), 81–109 (2006)
11. Zahariev, E.V.: Numerical multibody system dynamics, rigid and flexible systems. Lambert Academic Publishing, GmbH & Co. KG (2012)

Printed in the United States
By Bookmasters